Azomethine ylide mediated addition reaction

Tushar Sutariya
Balvantsingh Labana
Bhagyashree Parmar

Azomethine ylide mediated 1,3 dipolar cycloaddition reaction

LAP LAMBERT Academic Publishing

Publisher:
LAP LAMBERT Academic Publishing
is a trademark of
International Book Market Service Ltd., member of OmniScriptum Publishing Group
17 Meldrum Street, Beau Bassin 71504, Mauritius

Printed at: see last page
ISBN: 978-620-0-32306-4

Chemistry of azomethine ylide mediated 1,3 dipolar cycloaddition reaction and its scope

By

Dr. Tushar R. Sutariya

Dr. Balvantsingh M. Labana

Dr. Bhagyashri D. Parmar

Preface

This book is helpful for post graduate level students and researchers of Organic chemistry to provide them detailed information and knowledge regarded 1,3 dipolar cycloaddition reaction via azomethine ylide.

This book contains literature data of previous work done on azomethine ylide mediated 1,3 dipolar cycloaddition reaction and based on that synthesis of pyrrolo-indole and isoquinolines. Synthesized compounds have been identified by spectral analysis using 2D NMR, DQF-COSY and NOESY, and single crystal X–ray diffraction studies their biological essay to analyse their potency and application.

I am especially thankful to my research guide Prof. N. J. Parmar for his continues support and invaluable guidance during my researcher work. I express my sincere thanks to Department of Chemistry, Sardar Patel University, V. V. Nagar for providing facilities to perform the research work mentioned in this book.

ABBREVIATIONS AND SYMBOLS

$\triangle$	Thermal
%	Percentage
μg	Microgram
μL	Micro litter
1,3-DC	1,3-Dipolar cycloaddition
APT	Attached proton test
Ar	Aromatic
Bn	Benzyl
bp	Boiling point
br	Broad
C	Carbon
°C	Degrees centigrade
Cal	Calculated
CCDC	Cambridge crystallographic data centre
CDCl$_3$	Deuterated chloroform
dd	Doublet of doublets
ddd	Doublet of doublet of doublets
de	Diastereomeric excess
DEPT	Distortionless enhancement by polarization transfer
DKHDA	Domino Knoevenagel–*hetero*–Diels–Alder
DMF	N,N-Dimethylformamide
DMSO	Dimethyl sulphoxide
DMSO-*d$_6$*	Deuterated dimethyl sulphoxide
DQF–COSY	Double quantum filtered-correlation spectroscopy
dr	Diastereomeric ratio
dt	Doublet of triplet
DW	Distilled water
EA	Ethyl acetate
Et	Ethyl
Et$_2$O	Diethyl ether
Fig	Figure
FRAP	Ferric reducing antioxidant power
FT–IR	Fourier transform-infrared spectroscopy
g	Gram
h	Hour
Hz	Hertz
hν	Photochemical
IL	Ionic liquid
J	Coupling constant
K	Kelvin

3

m	Multiplet
M.P.	Melting point
MCRs	Multi component reactions
MIC	Minimal inhibitory concentration
MTB	*Mycobacterium tuberculosis*
MTCC	Micro type culture collection
MW	Microwave
NCCLS	National committee for clinical laboratory standards
NMR	Nuclear magnetic resonance
NOE	Nuclear overhauser effect
NOESY	Nuclear overhauser effect spectroscopy
NTM	Nontuberculous mycobacterium
ORTEP	Oak ridge thermal ellipsoid plot program
ppm	Parts per million (10^{-6})
PTC	Phase transfer catalyst
ROS	Reactive oxygen species
rt	Room temperature
s	Singlet
t	Triplet
TB	*Tubercle bacillus*
TBA–HS	Tetrabutylammonium hydrogen sulfate
TEAA	Triethyl ammoniumacetate
THF	Tetra hydro Furan
THIQ	Tetrahydroisoquinoline
THIQs	Tetrahydroisoquinolines
TLC	Thin layer chromatography
TS	Transition state
δ	Chemical shift
λ	Wave length
ν	Stretching vibrations
sex	Sextet
Quin.	Quintet
r.t.	Room temperature
cm^{-1}	Wave number
^{1}H NMR	Proton resonance spectrum
2D	Two dimensional
^{13}C NMR	^{13}Carbon resonance spectrum

Index

Chapter	1	**1,3 Dipolar Cycloaddition Reaction**	
	1.1	Introduction of 1,3 Dipolar Cycloaddition Reaction	7
	1.2	1,3 Dipoles	7
	1.3	Dipolarophiles	9
	1.4	References	9
Chapter	**2**	**Introduction of Azomethine ylide and its Scope**	
	2.1	Introduction of Azomethine ylide	11
	2.2	Introduction of some heterocycles moiety	14
	2.3	References	22
Chapter	**3**	**Experimental Part**	
	3.1	Synthesis	27
	3.2	Biological Screening test	35
	3.3	References	39
Chapter	**4**	**Results and Discuss**	
	4.1	Mechanism of 1,3 DC Reaction	41
	4.2	Evaluation of spectroscopic data	42
	4.3	Single crystal X-ray data	43
	4.4	Biological evaluation	44
	4.5	Characterization	51
	4.6	References	74

1,3 Dipolar Cycloaddition Reaction

1.1 Introduction

The 1,3-dipolar cycloaddition is a chemical reaction between a dipolarophile (alkene or alkyne) and 1,3-dipole to form a five-membered heterocyclic ring. There are mainly two type of 1,3-dipolar cycloaddition reaction, intermolecular 1,3-dipolar cycloaddition and intramolecular 1,3-dipolar cycloaddition (Fig. 1). The reaction is highly stereoselective reaction i.e. enantioselective and diastereoselective and can be used to synthesize five-member ring of central skeletons of several natural products, pharmaceuticals, organocatalysts, and other compounds of biological interest.[1]

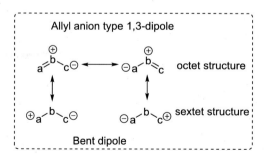

Fig. 1. General aspect of Intermolecular and intramolecular 1,3- dipolar cycloaddition reaction.

1.2 1, 3-Dipoles

1,3-Dipoles are two types.[2]

1) Allyl anion type 1,3-Dipoles, having a 4π-delocalized electron-system that has one emptied and two filled orbitals. Possible resonance structures of Allyl anion type a 1,3-dipole is given below.

Examples of allyl anion type 1,3-dipoles.

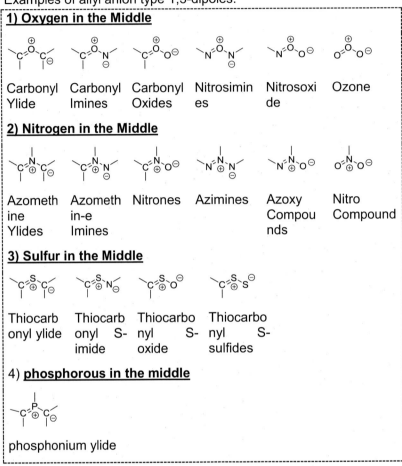

1) Oxygen in the Middle					
Carbonyl Ylide	Carbonyl Imines	Carbonyl Oxides	Nitrosimines	Nitrosoxide	Ozone

2) Nitrogen in the Middle					
Azomethine Ylides	Azomethin-e Imines	Nitrones	Azimines	Azoxy Compounds	Nitro Compound

3) Sulfur in the Middle			
Thiocarbonyl ylide	Thiocarbonyl S-imide	Thiocarbonyl S-oxide	Thiocarbonyl S-sulfides

4) **phosphorous in the middle**

phosphonium ylide

2) Propargyl or allenyl anion type containing extra π orbital which is orthogonal to molecular orbital (MO). Possible resonance structures of Propargyl or allenyl anion type 1,3-dipole is given below.

Allenyl anion type 1,3-dipole

Linear dipole

Examples of Propargyl/allenyl anion type 1,3-dipoles.

Diazonium Betaines			Nitrillium Betaines		
$N\equiv N-C$ ⊕ ⊖/	$N\equiv N-N$ ⊕ ⊖	$N\equiv N-O$ ⊕ ⊖	$-C\equiv N-O$ ⊕ ⊖	$-C\equiv N-N$ ⊕ ⊖	$-C\equiv N-C$ ⊕ ⊖/
Diazo-alkanes	Azides	Nitrous Oxide	Nitrile Oxides	Nitrile Imines	Nitrile Ylides

1.3 Dipolarophile

An alkene or alkyne which react with 1,3 dipole in cycloaddition reaction is called dipolarophile. Examples include α, β-unsaturated carbonyl compounds (**1**), ketones (**2**), allylic alcohols (**3**), allylic halides (**4**), alkynes (**5**), vinylic ethers (**6**), vinylic esters (**7**) and imines (**8**) (Fig 2).[3]

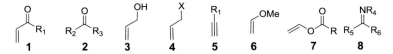

Fig 2 Dipolarophiles.

1.4 References:

1. (a) Cheng, Y.; Huang, Z. -T.; Wang, M.-X. *Curr. Org. Chem.* **2004**, 8, 325; (b) Notz, W.; Tanaka, F.; Barbas III, C. F. *Acc. Chem. Res.* **2004**, 37, 580; (c) Felpin, F.-X.; Lebreton, J. *Eur. J. Org. Chem.* **2003**, 3693; (d) Pearson, W. H.; Stoy, P. *Synlett* **2003**, 903. (e) Pearson, W. H. *Pure Appl. Chem.* **2002**, 74, 1339; (f) Pharmaceuticals, Vol. 1–4 (Ed.: McGuire, J. L.), Wiley-VCH, Weinheim, **2000**.

2. (a) Smith, M. B.; March, J. *March's Advanced Organic Chemistry*; 5 ed.; John Wiley & Sons, Inc.: New York, **2001**; (b) Stanley, L. M.; Sibi, M. P. *Chem. Rev.* **2008**, *108,* 2887.

3. (a) Atkinson, R. S. *Stereoselective Synthesis*, Wiley and Sons: England, **1995**, 173, 231; (b) Padwa, A.; Weingarten, M. D. *Chem. Rev.* **1996**, *96*, 223.

Chapter–2

Introduction of Azomethine ylide and its Scope

Introduction of Azomethine ylide and its Scope

2.1 Introduction of Azomethine ylide

The azomethine ylide is a planar allyl anion type 1,3-dipole, which contains a central nitrogen atom attached to two terminal carbons (Fig 1).[1] Except few cases,[2] the dipole is a very reactive, unstable and short-lived. Due to this, its stable precursor—derived—*in situ* and generated form is often trapped in the 1,3-DC reaction affording nitrogen containing five–membered heterocycles such as pyrrolidines or pyrroles. Over the past few decades, the strategy has become popular and come out with a variety of methods.

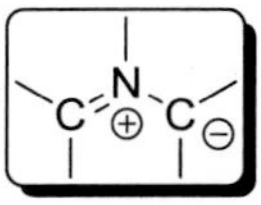

Fig 1 The general structure of azomethine ylides.

Generating four new chiral centers, 1,3-DC reaction gives various stereo-isomers involving azomethine ylide and so leading to the high levels of stereo-selectivity. Out of the four possible geometries of the azomethine ylide which determines the stereo-chemical outcome, W-and U-shaped geometries give rise to a 2,5-*cis*-disubstituted cyclic amine product *via* suprafacial transformation, while two S-shaped geometries to a 2,5-*trans*-disubstituted product (Fig 2). Isomerization of the ylide leads to the formation of a mixture. The 2,5-*trans*-disubstituted pyrrolidines are generally preferred with S-shape geometry, resulted from the aldehydes and unhindered secondary *α*-amino esters.[3]

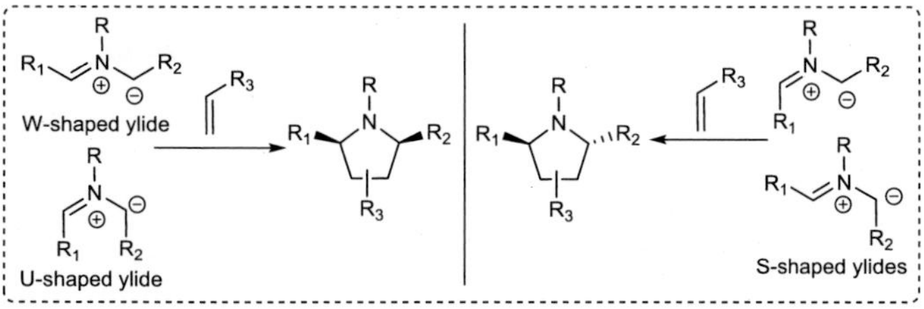

Fig 2. The different geometries of the azomethine ylide.

A large number of methods have been developed for the generation of azomethine ylide such as the proton abstraction from the imine derivatives of α-amino acids,[4] the thermolysis and photolysis of aziridines,[5] the condensation of aldehyde with secondary amine,[6] the decomposition of N-alkyl-N-methoxymethyl-N-(trimethylsilyl) methylamines,[7] desilylation of various α-amino silane derivatives,[8] decarboxylative condensation of amino acids,[9] deprotonation of iminium salts,[10] and others.[11] Among them, first three are most common procedures.

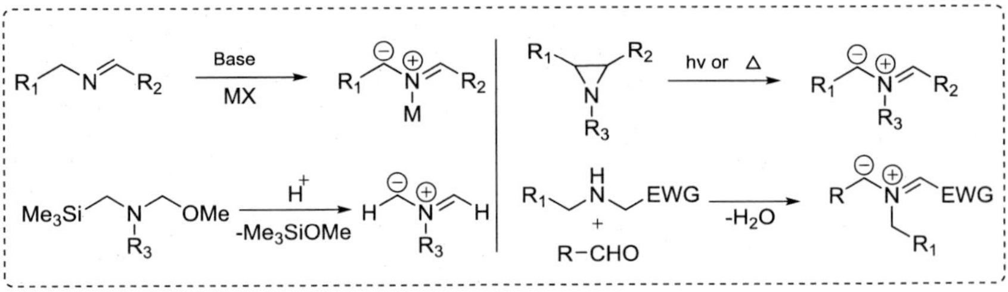

2.1.1 1, 3-Dipolar Cycloaddition reactions of azomethine ylides.

1,3-DC reaction occurring *via* the *in situ* generated azomethine ylide represents a significant transformation in the heterocyclic preparations.[12] It allows the formation of two bonds and up to four stereogenic centers in a single operation (Fig 3).[13] The reaction can be performed in two different versions; the intermolecular and intramolecular. In first type, intermolecular 1,3- dipolar cycloaddition, the azomethine ylide and dipolarophile (alkene or alkyne) are present into different molecule to react. In the second type, intramolecular 1,3- dipolar cycloaddition, both azomethine ylide and dipolarophile are present within the same molecule, resulting into the considerable complexity.[14]

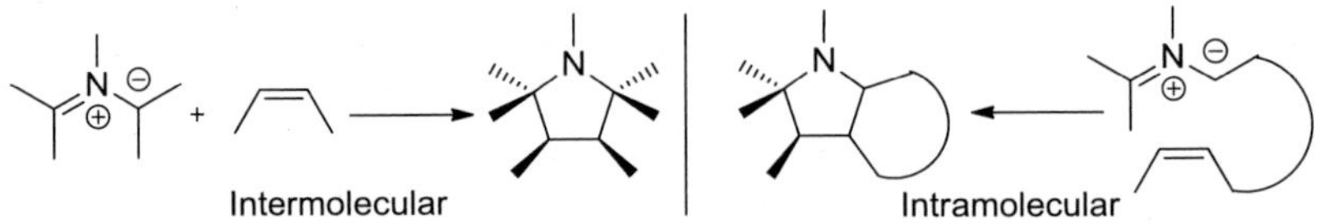

Intermolecular | Intramolecular

Fig 3. The cycloaddition reaction of the azomethine ylide.

2.1.2 Azomethine ylide in the synthesis of natural products.

The chemistry of azomethine ylide has been used in the total synthesis of number of natural products, as described below

Confalone and Earl[15] described the synthesis of alkaloid lycorenine **5** by heating typical aldehyde **1** with amino acid ester derivative **2**. Here, the *in situ* generated azomethine ylide **3** underwent smoothly [3+2] intramolecular cycloaddition forming octahydroindole ring system **4**.

Banwell *et al.*[16] prepared lamellarin K **11** *via* azomethine ylide generated *in situ* from the dihydroisoquinolinium salt **9**, through [3+2] cycloaddition reaction. The de-isopropylation of the product, pyrrole, **10**, takes place into marine alkaloid lamellarin K **11**.

Pearson and Lovering[17] achieved the synthesis of (±)-crinine **16** in eight steps, with overall 20% of the yield, as a single stereoisomer of the perhydroindole, from **12**.

This chapter describes one-pot synthesis of some N–fused indole and isoquinoline derivatives, *via* 1,3-DC reaction of azomethine ylide; generated *in situ* from N–allylated indole–carbaldehyde and secondary acyclic as well as cyclic amines, in an ionic liquid, TEAA. Stereochemical outcome of the reaction was established by 2D NMR experiments; DQF–COSY and NOESY, and single crystal X–ray diffraction data. Compounds were also screened *in vitro* for determining their antimicrobial, antituberculosis, antioxidant, and antiproliferative activities.

2.2 Introduction of Some heterocyclic Moiety

2.2.1 N-Fused pyrrolo-indole

Indole fusion to a five–membered ring *via* its side 'a', leads to the generation of a versatile fused ring system called N–fused pyrrolo–indole, in which nitrogen is common to both the rings.

N–Fused indole–heterocycles are integral structural core units of many naturally occurring plant species, in medicines and pharmaceuticals,[18] constituting an important class of bioprofiles. Following are notable examples of class of these heterocycles.

Alkaloid Mitomycin C **17** is known for its potent anti-tumoral and antibacterial activities[19], containing indole-pyrallidine fusion. Cytotoxicity of cyclopropamitosenes that contain a specific fusion pyrrolo[1,2-*a*]indole is well established by bacterial cell division and DNA alkylation.[20] Flinderoles A-C **18,** on the other hand, found in initial anti-malarial natural product extract, has selective anti-malarial activity against the *P. falciparum parasite* in general, while flinderole C in particular most active with IC$_{50}$ value 150 nM. Indigotica, as a plant species widely cultivated in China and East Asia, used as a traditional medicine for treatment of viral diseases such as influenza, viral pneumonia, mumps, and hepatitis.

Further investigations by Chen and co-workers led to the isolation of Isatisne A[22] **20,** with potent antiviral compounds.

Mitomycin C (17) **Flinderole A-C (18)** **Yuremamine (19)** **Isatisine A (20)**

Next, a member of alkaloids, Yuremamine **19,** is a new phytoindole that could be isolated from the stem bark of Mimosa hostiles, as potent hallucinogenic agent[23]. Many reports on synthesis of class of these heterocycles have appeared in the literature. Cited below are some key reports highlighting the status of the present work.

Min Shi *et al.* reported[24] a thermally-induced [3+2] ring-opening and cyclization reactions, combining aniline-tethered alkylidenecyclopropanes **21** with aldehydes **22,** and synthesized pyrrolo[1,2-a]indoles **23.**

Beccalli *et al.* reacted[25] N-allyl-2-indole-carbaldehydes **24** with N-benzylhydroxylamine hydrochloride **25,** and afforded indole-fused five–membered **26** and six–membered **27** heterocycles, *via* 1,3-DC reaction of *in situ* generated nitrone intermediate, regio-selectively.

Nobuharu *et al.* reported[26] effective protocol for N-fused polycyclic indole derivatives **29**, wherein a metal-containing azomethine ylide was generated *in situ* from *N*-(*O*-alkynylphenyl) imine derivatives **28** and W (CO)$_5$(L), for [3 +2] cycloaddition reaction.

Martinelli *et al.*[27] used glutaric anhydride **30** against bis-(trimethysilyl) acetylene in the presence of AlCl$_3$, and afforded 7-trimethysilyl-5-oxo-hept-6–ynoic acid. It gave N-benzyl-N-(7-trimethysilyl-5-oxohept–6-yn-l-oyl)-alanine methyl ester **32** with methyl proline **31,** in condensation reaction, in presence of 2-chloro-4,6-dimethoxy-1,3,5-triazine (CDMT). Finally [3+2] cycloaddition reaction followed by CO$_2$ removal gave substituted pyrrolo[1,2-*a*]indole **33**.

Zecchi *et al.* reported[28] polycyclic indole-heterocycles **36** employing reaction of N-(cyclohex-2-enyl) indole-2-carbaldehyde **34** with *N*-benzylhydroxylamine hydrochloride **35,** generating *in situ* nitrone intermediate that reacted intramolecularly with cyclohexynyl type dipolarophile.

2.2.2 N-fused Pyrrolo-isoquinolines

Isoquinoline fusion with pyrrolidine through its side *'a'*, and in which nitrogen is a common element of both the rings leads to generate N-fused pyrrolo-isoquinoline framework.

These framwork represents important structural motifs existing in many alkaloids as core units, and displaying significant biological and pharmacological properties. Isoquinolines with substitent at its first position are potent pharmacological agents, and so are targtes of many synthetic and bi-chemists,[29] revealing diverse biactivities such as agonism of the β-adrenoceptor,[30] antagonism of the D1 and NMDA receptors,[31] inhibition of α-glucosidase[32] and neurotoxicity associated with Parkinson's disease.[33] A majority of heterocycles have a chiral center at the C–1 position of the isoquinoline core, and stereoisomers at this position exhibit very different activities.

(S)-(-)-Troline **37**   (R)-(+)-Crispine A **38**   (R)-(+)-Oleracein E **39**   Erythrocarine **40**

(+)- Jamtine **41**   Annosqualine **42**   (-)-Cryptaustoline **43**

Alkaloids incorporating tricyclic pyrrolo[2,1-*a*]isoquinoline core structure are many, which are known for their pharmacological properties. Notable examples include *Portulaca oleracea L.* (Portulacaceae) containing alkaloid (R)-(+)-oleracein E **39**, widely distributed in weed. It is used as a folk medicine in many countries, exhibiting a wide range of pharmacological effects, including analgesic, anti-inflammatory,[34] skeletal musclerelaxant[35] and wound-healing[36] activities. Other alkaloids such as (S)-(-)-trolline **37**, (R)-(+)-crispine A **38**,[37] erythrocarine **40**, (+)-jamtine **41**,[39] annosqualine **42**[40] and (-)-cryptaustoline **43**,[41] also have the tricyclic pyrrolo[2,1-*a*]isoquinoline.

Several substituted pyrrol-annulated isoquinoline frameworks have been appeared synthetically feasible in view of searching for the new, medicinally important compounds. Further, a variety of reaction conditions such as MW-irradiations, surfactant, water, solid-support, and solvent reflux etc., have been reported to govern 1,3-DC reaction. Some selected methods are discussed below.

Marx *et al.* synthesized polycyclic lactams **46** by solvent-mediated and polymer-supported intramolecular cyclization, from readily available aldehydes and 2° amines.[42]

Yan *et al.* synthesized[43] a series of complex spiro[indoline-3,3'-pyrrolo[1,2-a]quinolines] **49**, regio–and diastereoselectively, by combining N-phenacylquinolinium bromides **47** with 3-phenacylideneoxindoles **48,** *via* triethylamine base-catalyzed [3+2] cycloaddition reaction in ethanol, which are thermodynamically stable trans isomers.

Carrillo *et al* reported[44] isoquinolinium methylides **50,** used as efficient azomethine ylide participating in [3+2] cycloaddtion reaction to the formation of pyrrolo-isoquinoline **53** in good yields, with high diastereo–and enantio–selectivities.

Dumitrascu *et al.*[45] synthesized pyrrolo[2,1-a]isoquinoline derivatives **57** by one-pot three component reactions taking isoquinoline **54,** 2-bromoacetophenones **55** and nonsymmetrical acetylenic dipolarophiles **56,** in 1,2-epoxypropane as a solvent.

Faulkner *et al.* reported[46] coupling reaction between the iodacetate **58** and 3,4-dihyreo-6,7-dimethoxyisoquinoline **59** to afford an intermediate salt **60,** which finally gave substituted pyrrolo[2,1-a]isoquinoline **61,** in presence of hunig's base, *via* [3+2] cycloaddition reaction.

2.2.3 Pyrrolizidines

Pyrrolizidine unit represents a fusion between two five–membered rings, in which nitrogen is a bridge-head element, common in both the rings.

Isolation and identification of several pyrrolizidine alkaloids (PAs) could be possible so far, available from thousands of plant species, mostly Boraginaceae, Graminae, Leguminosae and Compositae. Insects including various species of butterflies and even animals such as amphibians are also good sources of these alkaloids.[47] Many of them caused hepatotoxic effects to both livestock and humans, and revealed carcinogenic, antineoplastic and genotoxic activities, primarily via DNA cross-linking.[48] Pyrrolizidine alkaloids consist of a necine base attached through one or more ester linkages to a necic acid. The most common necine base is retronecine, with approximately 40% of all structurally elucidated pyrrolizidine alkaloids containing retronecine.[49]

lasiocarpine (62) europine (63) heliovicine (64)

lycopsamine (65) alexine (66) casuarine (67) hyacinthacine A$_2$ (68)

Lasiocarpine (**62**) and europine (**63**), which could be obtained from Heliotropium bovei, function as deterrents to insect feeding. In addition, they showed general antimicrobial and antifungal activities against fusarium moniliforme. [50] Heliovicine (**64**) and lycopsamine (**65**) could be isolated from the Heliotropium floridum and the Heliotropium megalanthum plants, respectively, and also showed anti-feedent activity against the Colorado potato beetle.[51] A number of PAs contain hydroxyls at multiple positions. Alexine[52] (**66**), available from Alexa canaracunensis, is a potential anti-HIV-1 agent. Casuarine (**67**), isolated from Casuarina equisetifolia and Eugenia jambolana, is effective in breast cancer treatment, bacterial infections, and diabetes[53]. Hyacinthacine A$_2$ (**68**), obtained from Muscari armeniacum, has strong amyloglucosidase and lactase inhibitory effects.[54]

Mishra *et al.* exploited[55] aldehyde **69**, derived from Morita–Baylis–Hillman adducts of acrylates, with methyl proline **70**, and obtained pyrido-fused pyrrolizidines **71**.

21

Burrell *et al.*[56] reported tricyclic pyrrolizidine **75** from acyclic precursors *via* cascade reaction involving condensation of an aldehyde **72** with a primary amine **73**, cyclization, (displacement of a halide), and then intramolecular 1,3-DC reaction through deprotonation or decarboxylation.

Jayashankaran *et al.* synthesized[57] dispiropyrrolizidino **79** ring systems. Azomethine ylides formed *via* decarboxylative step afforded typical products with dipolarophile 9-arylidine-fluorene **76**, in shorter reaction time, with higher yields.

2.3 References:

1. Husinec, S.; Savic, V. *Tetrahedron: Asymmetry* **2005**, *16*, 2047.
2. (a) Grigg, R.; Malone, J. F.; Mongkolaussavaratana, T.; Thianpatanagul, S. *Tetrahedron* **1989**, *45*, 3849; (b) Fleury, J. P.; Schoeni, J. P.; Clerin, D.; Fritz, H. *Helv. Chim. Acta* **1975**, *58*, 2018.
3. Coldham, I.; Hufton, R. *Chem. Rev.* **2005**, *105*, 2765.
4. Tsuge, O.; Kanemasa, S.; Yoshioka, M. *J. Org. Chem.* **1988**, *53*, 1384.
5. (a) Huisgen, R.; Cheer, W.; Huber, H. *J. Am. Chem. Soc.* **1967**, *89*, 1753; (b) DeShong, P.; Kell, D. A. *Tetrahedron Lett.* **1986**, *27*, 3979.
6. Panja, S. K.; Karmakar, P.; Chakraborty, J.; Ghosh, T.; Bandyopadhyay, C. *Tetrahedron Lett.* **2008**, *49*, 4397.
7. Harwood, L. M.; Vickers, R. J. In *Synthetic Applications of 1,3-Dipolar Cycloaddition Chemistry Toward Heterocycles and Natural Products*; Padwa, A.; Pearson, W. H., Eds.; John Wiley & Sons, Inc., New York, **2002**, vol 59, 170.
8. Carruthers, W.; Coldhanm, I. *Modern Methods of Organic Synthesis*, Cambridge Universty Press, 2004, 228.

9.   Döndaş, A. H.; Fishwick, C. W. G.; Grigg, R.; Kilner, C. *Tetrahedron* **2004**, *60*, 3473.

10. Kraus, G. A.; Nagy, J. O. *Tetrahedron Lett.* **1983**, *24*, 3427.

11. Pearson, W. H.; Stoy, P.; Mi, Y. *J. Org. Chem.* **2004**, *69*, 1919.

12. (a) Lown, J. W. In *1,3-Dipolar Cycloaddition Chemistry*; Padwa, A., Ed.; Wiley: New York, 1984; Vol. 1, p 653; (b) Vedejs, E. In *Advances in Cycloaddition*; Curran, D. P., Ed.; JAI Press: Greenwich, CN, 1988; Vol. 1, p 33; (c) Grigg, R. *Chem. Soc. Rev.* **1987**, *16*, 89.

13. Grigg, R.; Malone, J. F.; Mongkolaussavaratana, T.; Thianpatanagul, S. *Tetrahedron* **1989**, *45*, 3849.

14. (a) Wade, P. A. In *Comprehensive Organic Synthesis*; Trost, B. M., Fleming, I., Eds.; Pergamon: Oxford, U.K., **1991**; Vol. 4, p 1111; (b) Namboothiri, I. N. N.; Hassner, A. *Top. Curr. Chem.* **2001**, *216*, 1.

15. Confalone, P. N.; Earl, R. A. *Tetrahedron Lett.* **1986**, *27*, 2695.

16. Banwell, M.; Flynn, B.; Hockless, D. *Chem. Commun.* **1997**, 2259.

17. Pearson, W. H.; Lovering, F. E. *J. Org. Chem.* **1998**, *63*, 3607.

18. (a) Nakataubo, F.; Cocuzza, A. J.; Keeley, D. E.; Kishi, Y. *J. Am. Chem. Soc.*, **1977**, *99*, 4835; (b) Nakatsubo, F.; Fukuyama, T.; Cocuzza, A. J.; Kishi, Y. *J. Am. Chem. Soc.*, **1977**, *99*, 8115; (c) Kishi, Y. J.; *Nat. Prod.*, **1979**, *42*, 549; (d) Hao, L.; Pan, Y.; Wang, T.; Lin, M.; Chen, L.; Zhan, Z. P. *Adv. Synth. Catal.*, **2010**, *352*, 3215; (e) Wolkenberg, S. E.; Boger, D. L.; *Chem. Rev.*, **2002**, *02*, 2477; (f) Crich, D.; Banerjee, A. *Acc. Chem. Res.*, **2007**, *40*, 151.

19. (a) Crooke, S. T.; Prestayko, A. W. Academic Press, New York, **1981**, vol. *3*, p. 49; (b) Alcaro, S.; Ortuso, F.; Coleman, R. S. *J. Med. Chem.*, **2002**, *45*, 861; (c) Zein, N.; Solomon, W.; Colson, K. L.; Schroeder, D. R. *Biochemistry*, **1995**, *34*, 11591; (d) Remers, W. A.; Dorr, R. T. in In Alkaloids Chemical and Biological Perspectives, ed. Pelletier, S. W. John Wiley & Sons, New York, **1988**, vol. *6*, pp. 1–74; (e) Coleman, R. S.; Burk, C. H.; Navarro, A.; Brueggemeier, R. W.; Diaz-Cruz, E. S., *Org. Lett.*, **2002**, *4*, 3545.

20. (a) Verboom, W.; Reinhoudt, D. N.; Lammerink, B. H. M.; Orlemans, E. O. M.; Van Veggel, F. C. J. M.; Lelieveld, P., *Anti-Cancer Drug Des.*, **1987**, *2*, 271; (b) Verboom, W.; Orlemans, E. O. M.; Scheltinga, M. W.; Reinhoudt, D. N.; Lelieveld, P.; Fiebig, H. H.; Winterhalter, B. R.; Double, J. A.; Bibby, M. C. *J. Med. Chem.*, **1989**, *32*, 1612; (c) Maliepaard, M.; de Mol, N. J.; Janssen, L. H. M.; Hoogvliet, J. C.; van der Neut, W.; Verboom, W.; Reinhoudt, D. N. *J. Med. Chem.*, **1993**, *36*, 2091.

21. (a) Fernandez, L. S.; Jobling, M. F.; Andrews, K. T.; Avery, V. M. *Phytother. Res.*, **2008**, *22*, 1409; (b) Dethe, D. H.; Erande, R. D.; Ranjan, A. *J. Am. Chem. Soc.*, **2011**, *133*, 2864; (c) Fernandez, L. S.; Buchanan, M. S.; Carroll, A. R.; Feng, Y. J.; Quinn, R. J.; Avery, V. M. *Org. Lett.*, **2009**, *11*, 329; (d) Fernandez, L. S.; Sykes, M. L.; Andrews K. T.; Avery, V. M. *Int. J. Antimicrob. Agents*, **2010**, *36*, 275.

22. (a) Suneel Kumar, C. V.; Puranik, V. G.; Ramana, C. V., *Chem. Eur. J.*, **2012**, *18*, 9601; (b) Yin, Q.; You, S. L.; *Chem. Sci.*, **2011**, *2*, 1344; (c) Patel, P.; Ramana,

C. V. *Org. Biomol. Chem.*, **2011**, *9*, 7327; (d) Karadeolian, A.; Kerr, M. A. *Angew. Chem., Int. Ed.,* **2010**, *49*, 1133.

23. Vepsa¨la¨inen, J. J.; Auriola, S.; Tukiainen, M.; Ropponen, N.; Callaway, J. C.; *Planta Med.,* **2005**, *71*, 1053.
24. Chen, K.; Zhang, Z.; Weib, Y.; Shi, M. *Chem. Commun.,* **2012**, *48*, 7696.
25. Beccalli, E. M.; Broggini, G.; Rosa, C. L.; Passarella, D.; Pilati, T.; Terraneo, A.; Zecchi, G. *J. Org. Chem.* **2000**, *65*, 8924.
26. Kusama, H.; Takaya, J.; Iwasawa, N. *J. Am. Chem. Soc.* **2002**, *124*, 11592.
27. Hutchison, D. R.; Nayyar, N. K.; Martinelli, M. J. *Tetrahedron Lett.,* **1996**, *37*, 2887.
28. Beccalli, E. M.; Broggini, G.; Farina, A.; Malpezzi, L.; Terraneo, A.; Zecchi, G.; *Eur. J. Org. Chem.* **2002**, 2080.
29. Chrzanowska, M.; Rozwadowska, M. D. *Chem. Rev.* **2004**, *104*, 3341.
30. Nikulin, V. I.; Rakov, I. M.; De Los Angeles, J. E. ; Mehta, R. C.; Boyd, L. Y.; Feller, D. R.; Miller, D. D. *Bioorg. Med. Chem.* **2006**, *14*, 1684.
31. (a) Wanner, K. T.; Beer, H.; Höfner, G.; Ludwig, M. *Eur. J. Org. Chem.* **1998**, 2019; (b) Gao, M.; Kong, D.; Clearfield, A.; Zheng, Q. -H. *Bioorg. Med. Chem. Lett.* **2006**, *16*, 2229.
32. Takada, K.; Uehara, T.; Nakao, Y.; Matsunaga, S.; van Soest, R. W. M.; Fusetani, N. *J. Am. Chem. Soc.* **2004**, *126*, 187.
33. (a) Shinohara, T.; Takada, A.; Toda, J.; Terasawa, N.; Sano, T. *Heterocycles* **1997**, *46*, 555; (b) Nagatsu, T. *Neurosci. Res.* **1997**, *29*, 99.
34. Chan, K.; Islam, M. W.; Kamil, M.; Radhakrishnan, R.; Zakaria, M. N. M.; Habibullah, M.; Attas, A. *J. Ethnopharmacol.* **2000**, *73*, 445.
35. (a) Parry, O.; Marks, J. A.; Okwuasaba, F. *J. Ethnopharmacol.* **1993**, *40*, 187; (b) Parry, O.; Okwuasaba, F.; Ejike, C. *J. Ethnopharmacol.* **1987**, *19*, 247.
36. Rashed, A. N.; Afifi, F. U.; Disi, A. M. *J. Ethnopharmacol.* **2003**, *88*, 131.
37. Louafi, F.; Moreau, J.; Shahane, S.; Golhen, S.; Roisnel, T.; Sinbandhit, S.; Hurvois, J. –P. *J. Org. Chem.,* **2011**, *76*, 9720.
38. Shimizu, K.; Takimoto, M.; Mori, M. *Org. Lett.,* **2003**, *5*, 2323.
39. Simpkins, N. S.; Gill, C. D. *Org. Lett.,* **2003**, *5*, 535.
40. Shigehisa, H.; Takayama, J.; Honda, T. *Tetrahedron Lett.* **2006**, *47*, 7301.
41. Meyers, A. I.; Sielecki, T. M. *J. Am. Chem. Soc.,* **1991**, *113*, 2789.
42. Marx, M. A.; Grillot, A. L.; Louer, C. T.; Beaver, K. A.; Bartlett, P. A. *J. Am. Chem. Soc.* **1997**, *119*, 6153.
43. Wu, L.; Sun, J.; Yan, C. –G. *Org. Biomol. Chem.,* **2012**, *10*, 9452.
44. Fernandez, N.; Carrillo, L.; Vicario, J. L.; Dolores Badia, D.; Reyes, E. *Chem. Commun.,* **2011**, *47*, 12313.
45. Dumitrascu, F.; Georgescu, E.; Georgescu, F.; Popa, M. M.; Dumitrescu, D. *Molecules* **2013**, *18*, 2635.
46. Ridley, C. P.; Reddy M. V. R.; Rocha, G.; Bushman, F. D.; Faulkner, J. *Bioorg. Med. Chem.* **2002**, *10*, 3285.
47. (a) Culvenor CCJ (**1980**) In: Smith, R. L.; Bababunmi, E. A. (eds) Alkaloids and Human Disease. Toxicology in the Tropics. Taylor & Francis, London; (b)

Mattocks AR (**1986**) Chemistry and Toxicology of Pyrrolizidine Alkaloids. Academic Press, London; (c) Bull, L. B.; Culvenor. C. C. J.; Dick, A. T. (**1968**) The Pyrrolizidine Alkaloids. North-Holland, Amsterdam.

48. (a) Segall, H. J.; Wilson, D. W.; Lamé, M. W.; Morin, D.; Winter, C. K. (**1991**) In: Keeler RF, Tu AT (eds) Metabolism of Pyrrolizidine Alkaloids. Handbook of Natural Toxins, vol 6. Marcel Decker, New York; (b) Swick, R. A. *J Anim Sci*, **1984**, *58*, 1017.

49. Logie, C. G.; Grue, M. R.; Liddell, J. R. *Phytochemistry*, **1994**, *37*, 43.

50. Reina, M.; Mericli, A. H.; Cabrera, R.; Gonzalez-Coloma, A. *Phytochemistry*, **1995**, *38*, 355.

51. (a) El Nemr, A. *Tetrahedron*, **2000**, *56*, 8579; (b) Reina, M.; Gonzalez-Coloma, A.; Gutierrez, C.; Cabrera, R.; Henriquez, J.; Villarroel, L. *Phytochemistry*, **1997**, *46*, 845.

52. Usuki, H.; Toyo-oka, M.; Kanzaki, H.; Okuda, T.; Nitoda, T. *Bioorg. Med. Chem.*, **2009**, *17*, 7248.

53. Asano, N.; Nash, R. J.; Molyneux, R. J.; Fleet, G. W. J. *Tetrahedron: Asymmetry* 2000, 11, 1645.

54. Asano, N.; Kuroi, H.; Ikeda, K.; Kizu, H.; Kameda, Y.; Kato, A.; Adachi, I.; Watson, A. A.; Nash, R. J.; Fleet, G. W. J.; *Tetrahedron: Asymmetry*, **2000**, *11*, 1.

55. Mishra, A.; Rastogi, N.; Batra, S. *Tetrahedron*, **2012**, *68*, 2146.

56. Burrell, A. J. M.; Coldham, I.; Watson, L.; Oram, N.; Pilgram, C. D.; Martin, N. G. *J. Org. Chem.* **2009**, *74*, 2290.

57. Jayashankaran, J.; Durga, R.; Manian, R. S.; Raghunathan, R. *Tetrahedron Lett.* **2004**, *45*, 7303.

Chapter–3

Experimental Section

Experimental

Literature survey showed that, substituted and/N-fused pyrrolo-indole-heterocycles are of particular interest, and have appeared as promising bioactive heterocycles in the literature. Azomethine ylide−triggered 1,3-DC reaction has been the effective mean to incorporate a heterocycle into N−fused pyrrolo-indole framework, and the scope of the route can still be extended to many new biomolecules. It was planned to synthesize pyrrolo-indole and pyrrolo-isoquinoline derivatives, as part of this chapter work. Use of N−allyl−3−chloro−indole−2−carbaldehyde against a variety of α-amino acid esters; N−methyl-/N-ethyl-/N-benzyl-glycine esters/4-(2-aminoethyl) morpholine, as well as tetrahydroisoquinolines, in ionic liquid TEAA, allowed the synthesis of title compounds effectively, *via* 1,3-DC reaction. All synthesized compounds were characterized by mass, IR, ^1H NMR and ^{13}C NMR spectroscopy data. Further, all the compounds were screened for their antimicrobial, antituberculosis, antioxidant and antiproliferative activities.

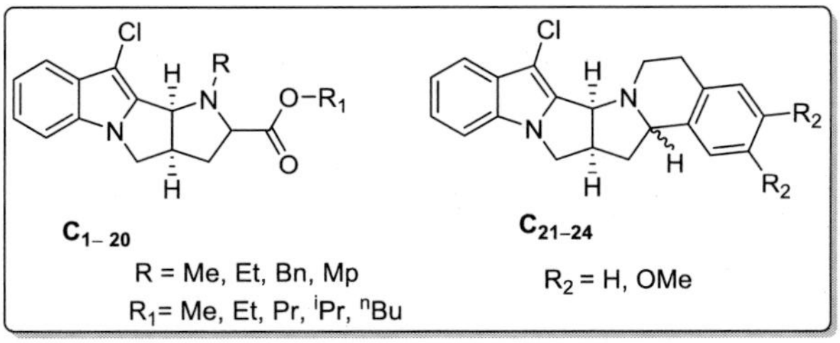

C_{1-20}
R = Me, Et, Bn, Mp
R_1= Me, Et, Pr, ^iPr, ^nBu

C_{21-24}
R_2 = H, OMe

Fig 3.1 General structural feature of compounds TC_{1-24}.

3.1 Synthesis

(A) Synthesis of pyrrolo-fused indoles and isoquinolines TC_{1-24}.

Steps given below describe synthesis of all title compounds.

a) Preparation of N-allyl−3-chloro-indole-2-carbaldehyde **1**.

  i. Preparation of (phenylglycine)-*o*-carboxylic acid.

  ii. Preparation of 3-chloroindole-2-carbaldehyde.

iii.Preparation of N-allyl 3-chloro-indole-2-carbaldehyde **1**.

b) Synthesis of secondary amines **2a-e** to **5a-e**.

c) Preparation of ionic liquid, TEAA.

d) Screening of the reaction conditions.

e) Synthesis of pyrrolo-fused indoles, and isoquinolines C_{1-24}.

a) **Synthesis of N-allyl 3-chloro-indole-2-carbaldehyde 1.**

(i) **Preparation of (phenylglycine)-o-carboxylic acid.[1]**

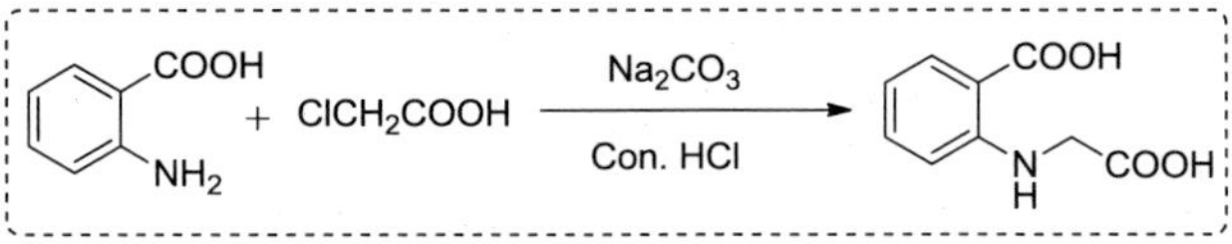

Scheme 1 Synthesis of (phenylglycine)-o-carboxylic acid.

General procedure: A mixture containing anthranilic acid (6.85 g, 50 mmol) and chloroacetic acid (4.73 g, 50 mmol) was dissolved in 50 mL water in a three–necked round–bottom flask–fitted with a reflux condenser, and heated with 50 mL aqueous sodium carbonate (100 mmol; 10.6 g) solution under reflux for 3 h. It was cooled down to room temperature, acidified slightly with concentrated HCl, and kept overnight. The solid crude product dropped out off the solution was filtered, washed by several water portions, and re–crystallized from hot water containing a little decolorizing carbon, yielding 70 % of the pure acid, with mp. 208–210 °C

(ii) **Preparation of 3-chloroindole-2-carbaldehyde.[2]**

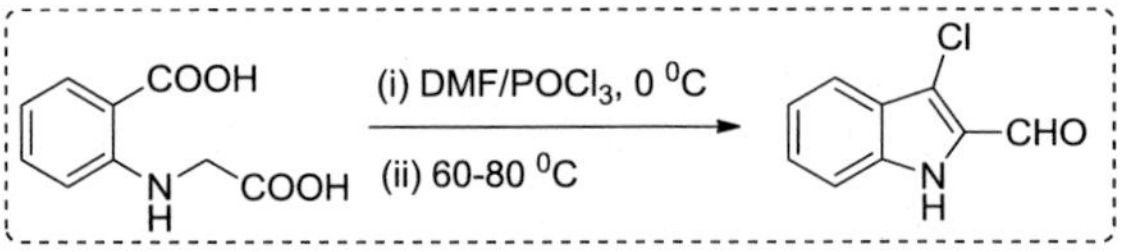

Scheme 2 Synthesis of 3-chloroindole-2-carbaldehyde.

General procedure: A three–necked round–bottom flask, equipped with thermometer, drying tube and mechanical stirrer, was charged with 9.3 ml of DMF. The resulted reaction content was cooled down to 0 ˚C, and was added drop–wise 5.5 mL of POCl₃ (60 mmol) and then (phenylglycine)-o-carboxylic acid (10 mmol, 1.95 g), with vigorous stirring. It was heated at 60–80 °C for 4–6 h, cooled down to r.t., and poured into crushed ice species with mechanical stirring. The solid precipitated out was filtered, and washed with several water portions to complete removal of acid, yielded 75 % of pure product, m.p. 172–174 °C.

iii) Synthesis of N-allyl 3-chloroindole-2-carbaldehyde 1.

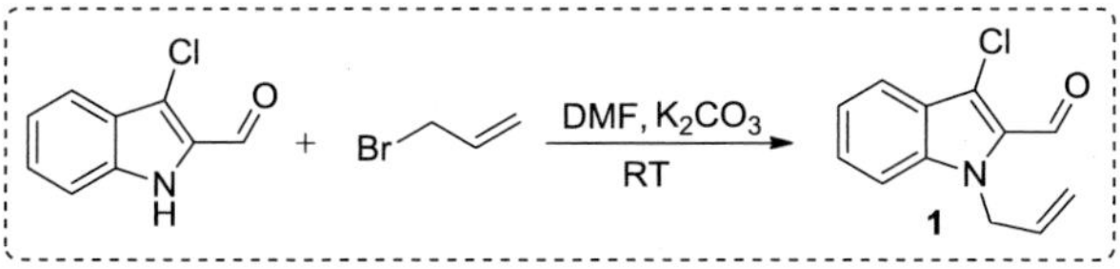

Scheme 3 Synthesis of N–allyl 3–chloroindole-2-carbaldehyde 1.

General procedure: A 10 mL of allyl bromide (15.0 mmol; 1.79 g) solution prepared in DMF was added drop-wise to a vigorously stirred mixture of 3-chloro-indole-2-carbaldehyde (10.0 mmol; 1.79 g) and anhydrous potassium carbonate (15.0 mmol; 2.06 g), suspended in DMF (25 mL). It was stirred well at r.t., and monitored periodically for starting components to disappear. After 10−12 h, it was then poured into ice species, with constant stirring. The end product was extracted with Et₂O (3×10ml) portions. Combined ether extract was dried over anhydrous Na₂SO₄, and evaporated to get residue of the product, in the 90–92 % range.

b) Synthesis of secondary amines 2a–e to 5a–e, and 6a–b*.

i) Synthesis of secondary amines 2a-e to 3a–e.

General procedure: A solution of corresponding ester ClCH₂COOR₁ (37 mmol; 4.01 g of methyl– or 4.53 g of ethyl– or 5.06 g of n-propyl/i-propyl– or 5.57 g of n-butylester) prepared in acetonitrile (10ml), was added drop-wise to a stirred suspension of amine hydrochloride RNH₂·HCl (37 mmol; 2.50 g of methyl– or 3.02 g of ethyl amine hydrochloride) at 15 °C with K₂CO₃ (74 mmole; 10.2 g) in acetonitrile (20ml) at r.t., with vigorously stirring for 10 h. The insoluble salts appeared were filtered off, washed with acetonitrile, and subjected to solvent evaporation under vacuum. It gave a residue of colorless liquid product, which can be used for next step without further purification.

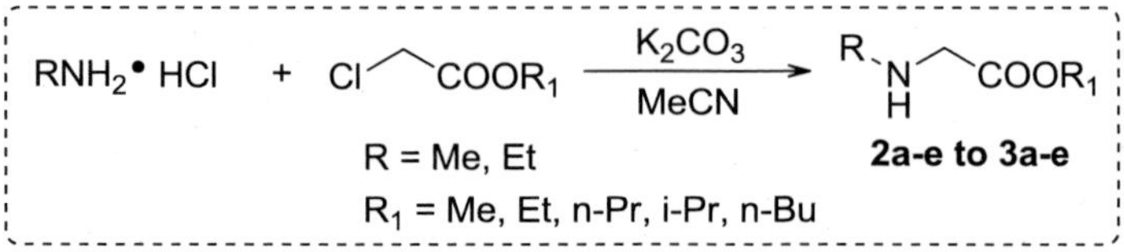

$$RNH_2 \cdot HCl \; + \; Cl \diagup COOR_1 \; \xrightarrow[\text{MeCN}]{K_2CO_3} \; R \diagdown_{\underset{H}{N}} \diagup COOR_1$$

R = Me, Et   **2a–e to 3a–e**

R_1 = Me, Et, n-Pr, i-Pr, n-Bu

Scheme 4 Synthesis of secondary amines **2a–e** to **3a–e**

Table 1 Synthesis of various amino acid esters $R_2NHCH_2COOR_1$ **2a–e** to **3a–e**.

Ester	R	R_1	Yield %	B. P. °C*
2a	Me	Me	89	116—119
2b	Me	Et	92	137—139
2c	Me	n-Pr	89	148—152
2d	Me	i-Pr	87	132—134
2e	Me	n-Bu	90	164—167
3a	Et	Me	91	137—140
3b	Et	Et	88	156—158
3c	Et	n-Pr	92	166—169
3d	Et	i-Pr	88	158—160
3e	Et	n-Bu	89	174—176

*Uncorrected, **2a—b** and **3a—b** are reported

ii) Synthesis of secondary amines 4a–e to 5a–e

General procedure: A corresponding chloroacetic acid ester $ClCH_2COOR_1$ (47 mmol; 5.06 g of methyl– or 5.72 g of ethyl– or 6.37 g of n-propyl/i-propyl– or 7.03 g of n-butyl ester), prepared in acetonitrile (10ml), was added drop wise to a stirred solution of amine (47 mmol; 5.00 g of benzyl amine or 6.10 g of 4-(2-aminoethyl)morpholine) and TEA (triethyl amine) (71 mmole; 9.88 ml) in acetonitrile (20ml), at r.t., with vigorous stirring for 4 h. The TEA·HCl salt precipitate dropped out of the solution were filtered, and washed with 5 mL acetonitrile portions. The filtrate along with acetonitrile washings was then evaporated to dry under vacuum to receive residue as colorless liquid product requiring no further purification.

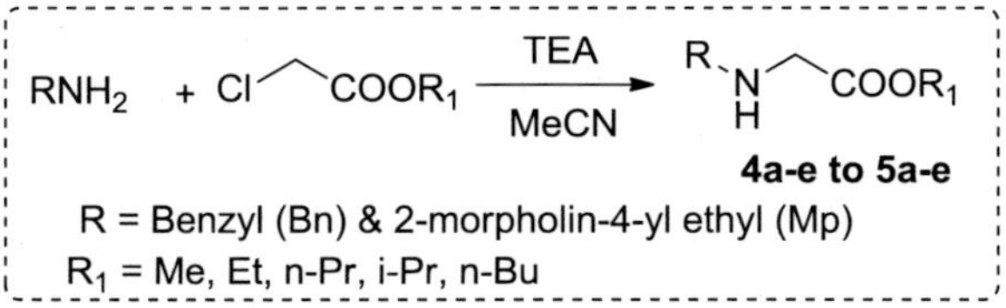

$$RNH_2 \; + \; Cl\!-\!COOR_1 \; \xrightarrow[\text{MeCN}]{\text{TEA}} \; R\!-\!\underset{H}{N}\!-\!COOR_1$$

4a-e to 5a-e

R = Benzyl (Bn) & 2-morpholin-4-yl ethyl (Mp)
R_1 = Me, Et, n-Pr, i-Pr, n-Bu

Scheme 5 Synthesis of secondary amines **4a–e** to **5a–e**.

Table 2 Synthesis of various amino acid esters $R_2NHCH_2COOR_1$ **4a–e** to **5a–e**.

Amino esters	R	R_1	Yield %	B. P. °C*
4a	Bn	Me	92	254—256
4b	Bn	Et	94	275—277
4c	Bn	n-Pr	93	286—288
4d	Bn	i-Pr	91	280—282
4e	Bn	n-Bu	92	302—304
5a	Mp	Me	90	272—276
5b	Mp	Et	88	284—286
5c	Mp	n-Pr	86	300—302
5d	Mp	i-Pr	87	292—294
5e	Mp	n-Bu	89	312—316

*Uncorrected, **4a—b** are reported

iii) **Cyclic secondary amine 6a–b.**

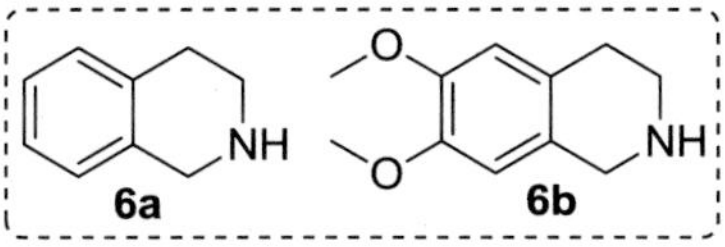

6a   6b

c) **Preparation of ionic liquid.**[3a]

Ionic liquid triethylammonium acetate (TEAA) has been prepared by the addition of 86.03 mL of glacial acetic acid (75 mmol, 4.50 g) drop-wise to 139.4 mL of triethylamine (50 mmol, 5.05 g), in a round bottom flask equipped with reflux condenser, within 1 h at 70 °C, and then heating of the resulted mixture at 80 °C for 2 h. Dried in high vacuum at 80 °C, the reaction mass yielded 96% of TEAA.

d)   Screening of the reaction conditions[3b]

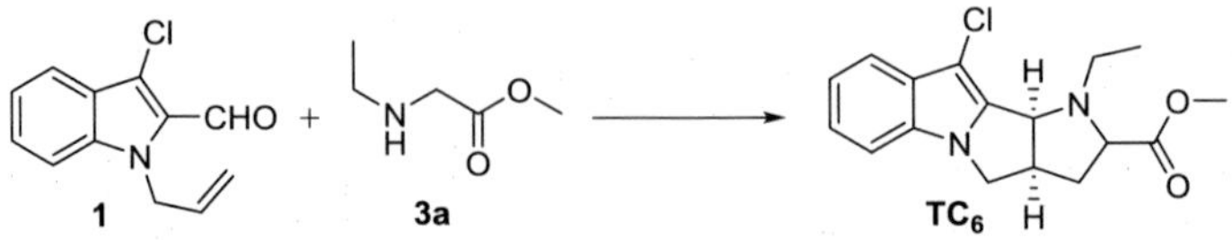

Scheme 6 A model reaction to optimize the reaction conditions

Table 3 Screening of the reaction conditions,

Entry	Solvent (reflux)	Catalyst	Temp (°C)	Time (h)	Yield (%)
1	MeOH	—	reflux	24	Trace
2	ACN	—	reflux	10	40
3	toluene	—	reflux	6.0	68
4	toluene	Na$_2$SO$_4$	reflux	6.0	72
5	xylene	—	reflux	5.5	70
6	xylene	Na$_2$SO$_4$	reflux	5.0	75
7	[A]	[A]	100	4.0	70
8	TEAA	TEAA	80	3.5	78
9	TEAA	TEAA	100	2.5	85

[A]solvent free

Out of various reaction conditions examined, the model reaction between **1** and **3a** was favoured in TEAA at 100 °C (entry 9). Above 100 °C, however a very little improvement was seen. This method was then generalised to prepare other products (Scheme 7, Table 2).

e)   Synthesis of pyrrolo-fused indoles and isoquinolines TC$_{1-24}$

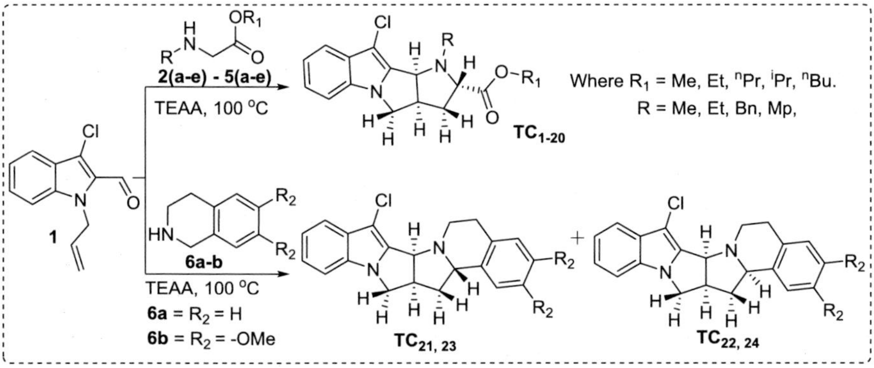

Scheme 7: Synthesis of pyrrolo-fused indoles and isoquinolines.

General procedure: A mixture of aldehyde **1** (4 mmol; 0.88 g) and acyclic secondary amine **2–5(a–e)** (4 mmol; 0.41 g of methyl— or 0.47 g of ethyl— or 0.52 g of n-propyl—/i-propyl— or 0.58 g of n-butyl N-methyl glycinate; 0.47 g of methyl— or 0.52 g of ethyl— or 0.58 g of n-propyl—/i-propyl— or 0.64 g of n-butyl N-ethyl glycinate; 0.72 g of methyl– or 0.77 g of ethyl— or 0.83 g of n-propyl—/i-propyl— or 0.88 g of butyl N—benzyl glycinate; 0.81 g of methyl– or 0.86 g of ethyl— or 0.92 g of n-propyl—/i-propyl— or 0.98 g of n-butyl N—(2-morpholinoethyl) glycinate) or isoquinoline **6a-b** (3 mmol; 0.53 g of 6a or 0.77 g of 6b) in 2 mL of ionic liquid TEAA in a round bottom flask, was heated at 100 °C and progress of the reaction was monitored by TLC. After the reaction was complete, the reaction mass was cooled down to r.t., and then poured into ice species. The oily product formed underwent emulsification with water, and hence required Et_2O (3X10 mL) portions to extract. Combined ether extract portions yielded quantitative amounts of crude product after evaporation of ether. Finally, the crude product was purified by column chromatography on silica gel using a 90:10 n-hexane-ethyl acetate mixture as an eluent. The yields recorded were in the 72-85 % range.

Table 4 Pyrrolo-fused indoles and isoquinolines TC_{1-24}.

Entry	Compound	R	R_1	R_2	Time(h)	Yield (%)	Melting point ($^\circ$C)
1	TC_1	Me	Me	—	2.75	82	118-120
2	TC_2	Me	Et	—	2.75	80	102-104
3	TC_3	Me	^nPr	—	3.0	77	106-108
4	TC_4	Me	^iPr	—	3.0	84	107-109
5	TC_5	Me	^nBu	—	3.5	77	92-94
6	TC_6	Et	Me	—	2.5	85	84-86
7	TC_7	Et	Et	—	2.75	82	92-94
8	TC_8	Et	^nPr	—	3.0	84	64-66
9	TC_9	Et	^iPr	—	3.0	78	62-63
10	TC_{10}	Et	^nBu	—	3.0	79	60-62
11	TC_{11}	Bn	Me	—	2.5	80	154-156
12	TC_{12}	Bn	Et	—	3.0	82	124-125
13	TC_{13}	Bn	^nPr	—	3.0	78	126-128
14	TC_{14}	Bn	^iPr	—	3.5	76	132-134
15	TC_{15}	Bn	^nBu	—	3.5	74	96-98
16	TC_{16}	Mp	Me	—	3.0	76	—
17	TC_{17}	Mp	Et	—	3.0	78	—
18	TC_{18}	Mp	^nPr	—	3.0	72	—
19	TC_{19}	Mp	^iPr	—	3.5	74	—
20	TC_{20}	Mp	^nBu	—	3.0	75	—
21	TC_{21}	—	—	H	2.75	45	118-120
22	TC_{22}	—	—	H	2.75	37	152-154
23	TC_{23}	—	—	OMe	3.0	46	162-164
24	TC_{24}	—	—	OMe	3.0	38	202-204

Aqueous part on the other hand, which was left after the ether extraction of oily products and containing TEAA, was treated further for the recovery ionic liquid. It was simply heated under reduced presser at 80 °C to remove the water. The recovered ionic liquid was used again for the same. TEAA can be recycled at least four-times without altering its efficiency.

3.2 Biological screening tests

(a) Antimicrobial activity

Procedure for performing the broth dilution method:

a) Three Gram-positive [MTCC (Micro Type Culture Collection), 1936 *Streptococcus pneumonia*, MTCC 449 *Clostridium tetani,* MTCC 441 *Bacillus subtilis*], three Gram-negative (MTCC 98 *Salmonella typhi,* MTCC 3906 *Vibrio cholerae,* MTCC 443 *Escherichia coli) bacteria,* and two fungi (MTCC 3008 *Aspergillus fumigates,* MTCC 227 *Candida albicans)* strains were procured from MTCC, Institute of Microbial Technology, Chandigarh, India, to investigate *in vitro* antimicrobial activity of all the heterocycles, comparing screening test results with that of standard reference drugs.

b) Inoculum size for test strain was adjusted to 10^8 CFUmL^{-1} (Colony Forming Unit per milliliter) by comparing the turbidity (turbidimetric method).

c) Mueller–Hinton broth was used as nutrient medium to grow and dilute the compound suspension for the test bacteria and Sabouraud-Dextrose broth used for fungal nutrition.

d) Ampicillin, Chloramphenicol, Ciprofloxacin, Gentamicin and Norfloxacin were used as standard antibacterial reference drugs, whereas Griseofulvin and Nystatin were standard antifungal reference drugs.

e) DMSO was used as diluents/vehicle to achieve a desired concentration of synthesized compounds. And a standard drug to test it upon a standard microbial strain.

f) Serial dilutions were prepared in primary and secondary screenings. A synthesized compound and a standard drug each was diluted to 2000 μgmL^{-1} concentration as a stock solution. In primary screening 1000, 500 and 250 μgmL^{-1} concentrations were taken. The active compounds if found active in the primary screening were further diluted to 200, 100, 125, 62.5, 50, 25, 12.5 and 6.250 μgmL^{-1} concentrations for secondary screening.

g) A Control tube containing no antibiotic is immediately subculture before incubation by spreading a loopful evenly over a quarter of the plate on a medium suitable for the growth in test organism. The tubes were incubated at 37 °C for 24 h in case of bacteria, and 48 h in case of fungi. The highest dilution or lowest concentration showing at least 99 % inhibition or preventing appearance of turbidity is considered as MIC (μgmL^{-1}). A set of tubes containing only seeded broth and the solvent controls were maintained under identical conditions so as to make sure that the solvent had no influence on strain growth. The results are much affected by the size of the inoculums. The test mixture should contain 10^8 CFUmL^{-1} organisms. The protocols were summarized and compared with standard antimicrobial and antifungal drugs in terms of MIC (μgmL^{-1}).

(b)   Antituberculosis activity

The MIC was used to evaluate the anti-tuberculosis activity. It is one of the non-automated *in vitro* bacterial susceptibility tests. This classic method yields a quantitative result for the amount of antimicrobial agents that is needed to inhibit growth of specific microorganisms. It is carried out in bottle.

Methods used for primary and secondary screening:

Each test compound was diluted to a 2000 μgmL^{-1}and was used as stock solution.

Primary screen: In primary screening, the 250 μgmL^{-1} concentration of synthesized compounds was taken. If compounds found active in primary screening, then they were tested further in a second set of dilution against all microorganisms.

Secondary screen: The compounds that found active in primary screening were similarly diluted to 500 μgmL^{-1}, 250 μgmL^{-1}, 200 μgmL^{-1}, 125 μgmL^{-1}, 100 μgmL^{-1}, 50 μgmL^{-1}, 25 μgmL^{-1}, 12.5 μgmL^{-1}, 6.25 μgmL^{-1}, 3.125 μgmL^{-1} and 1.5625 μgmL^{-1} concentrations.

A primary screening was conducted at 250 μgmL^{-1} against *M. tuberculosis* H37Rv following a Lowenstein-Jensen (L-J) MIC method.[4] Test compounds were added to liquid L-J medium and then media were sterilized by inspissation method. A culture of *M. tuberculosis* H37Rv grown on L-J medium was harvested in 0.85% saline in bijou bottles. DMSO was used as vehicle to get a desired concentration. These tubes were then incubated at 37 °C for 24 h followed by streaking of *M. tuberculosis* H37Rv (5 × 10^4 bacilli per tube). These tubes were then incubated at 37 °C. Growth of bacilli was seen after 12, 22, and finally 28 days incubation. Tubes having the compounds were compared with control tubes where medium alone was incubated with *M. tuberculosis* H37Rv. The concentration at which complete inhibition of colonies occurred was taken as active concentration of test compound. The standard strain *M. tuberculosis* H37Rv was tested with known drugs isoniazid and rifampicin. The screening test results are summarized as % inhibition relative to standard drugs isoniazid and rifampicin. Compounds effecting < 90 % inhibition in the primary screen were not evaluated further. Compounds showed at least 90 % inhibition in the primary screen was re-tested at lower concentration (MIC) in a L-J medium.

(c)   Antioxidant activity

The FRAP assay method, was used to measure the capacity of test compound to reduce ferric tripyridyl-s-triazine (Fe(III) TPTZ) complex to the ferrous tripyridyltriazine (Fe(II)-TPTZ) at low pH. The complex (Fe(II)-TPTZ) has an intensive blue color and can be monitored at λ_{max} 593 nm.

Reagents:

1) Acetate buffer, pH 3.6, 300 mmol/L; (3.1 g sodium acetate trihydrate and 16 mL glacial acetic acid per liter of buffer solution).

2) 10 mmol/L, 2,4,6-tripyridyl-s-triazine (TPTZ) (MW 312.34) solution in 40 mmol/L HCl.

3) 20 mmol/L, $FeCl_3 \cdot 6H_2O$ (MW 270.30) in distilled water (DW).

FRAP working solution: It was prepared by mixing 25 mL of acetate buffer (1), 2.5 mL of TPTZ and 2.5 mL of $FeCl_3 \cdot 6H_2O$ solutions. In the study, the working solution was always kept fresh.

Sample solution: 0.005 g in 25 mL DMF.

Standard solution: 1mmol of ascorbic acid (MW 176.13 g/mol) in 100 mL DW.

Procedure: A 400 μL of sample or 200 µL of a standard was mixed with 3.0 mL FRAP working solution, and incubated the resulted mixture at 37 °C for 10 min. Its absorbance was measured at 593 nm using respective blank solutions. The antioxidant activity is expressed as ascorbic acid equivalent (mmol/100 g).

Calculation:

FRAP value was calculated by following equation: [5]

$$\text{Ascorbic acid concentration (mM/100 g of sample)} = \frac{\Delta A_{593nm} \text{ of test sample}}{\Delta A_{593nm} \text{ of standard}} \times \frac{\text{Standard (mmol)}}{\text{Sample (mg)}} \times 100$$

(d) Antiproliferative activity:

The antiproliferative assay for antitumor activity was performed in 96-well plates using the National Cancer Institute (NCI) protocol.[6] As a model to study the anticancer activity of the synthetic compounds, six human solid tumor cell lines; A549 (non-small cell lung), HBL-100 (breast), HeLa (cervix), SW1573 (non-small cell lung), T-47D (breast), and WiDr (colon) were used. The *in vitro* antiproliferative activity was evaluated after 48 h of drug exposure using the sulforhodamine B (SRB) assay.[7] According to NCI protocol; only compounds soluble in DMSO at 40 mM were tested.

References:

1. Haller, H. L. *J. Ind. Eng. Chem.* **1922**, *14*, 1040.
2. Vattoly, J. M.; Paramasivan, T. P. *J. Org. Chem.* **1996**, *61*, 6523.
3. a) Verma, A. K.; Attri, P.; Chopra, V.; Tiwari, R. K.; Chandra, R. *Monatsh Chem.* **2008**, *139*, 1041.
   b) Sutariya, T. R,; Labana, B. M.; Parmar, N. J.; Rajni Kant, Gupta, V, K.; Plata G. B.; Padron, J. M. New. J. Chem., **2015**, 39, 2657.
4. Rattan, A. *Antimicrobials in Laboratory Medicine;* Churchill, B. I.; Livingstone, New Delhi, **2000**; 85.
5. (a) Benzie, I. F. F.; Strain, J. J. *Redox Rep.* **1997**, *3*, 233; (b) Benzie, I. F. F.; Strain, J. J. *Methods in Enzymology: Oxidants and Antioxidants Part A.* 1st Ed.; London: Academic Press Limited, **1999**; Vol. 299, p 15; (c) Choy, C. K. M.; Benzie, I. F. F.; Cho. P. *IOVS,* **2000**, *41*, 3293.
6. Monks, A.; Scudiero, D.; Skehan, P.; Shoemaker, R.; Paull, K.; Vistica, D.; Hose, C.; Langley, J.; Cronise, P.; Vaigro-Wolff, A.; Gray-Goodrich, M.; Campbell, H.; Mayo, J.; Boyd, M. *J. Natl. Cancer Inst.* **1991**, *83*, 757.
7. Skehan, P.; Storeng, P.; Scudeiro, D.; Monks, A.; McMahon, J.; Vistica, D.; Warren, J. T.; Bokesch, H.; Kenney, S.; Boyd, M. R. *J. Natl. Cancer Inst.* **1990**, *82*, 1107.

Chapter–4

Results and Discussion

4 Results and discussion

4.1 Mechanism of azomethine ylide mediated 1,3 DC reaction

A plausible mechanism of the 1,3 Dipolar cycloaddition reaction has been described in the below Scheme 1.

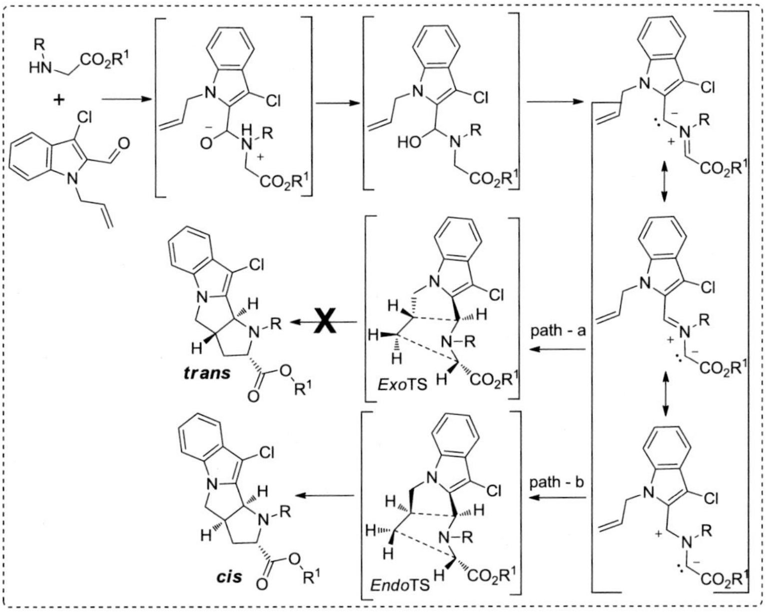

Scheme 1 A plausible mechanism of the 1, 3-DC reaction, synthesis of **TC$_{1-20}$**.

Addition of electron starts from the tethered-alkene terminal carbon on imine, which in turn follows the addition of the enolized ester on the other end of the alkene through semi bicyclic transition state, but it may force system to adopt the most favoured *cis*-fusion product. Stereoselectivity of the reaction is due to *exo* or *endo* attack of dipolarophile alkene on azomethine ylide dipole. The spectral data support the *cis*-fusion between central pyrrolidine rings in all heterocycles, which favours the *endo* transition state. In **TC$_{22}$ $_{\&}$ $_{24}$**, however, the *cis*-fusion might have involved a further isomerisation of the ylide.[1]

4.2   Evaluation of Spectroscopic data

The proposed structures of all heterocycles are fully agreed with observed mass, IR and NMR spectral data. A sharp IR band observed in the 1725–1735 cm^{-1} range in **TC$_{1-20}$** indicates the presence of ester carbonyl (–COO–) group. Other bands; one in the 3050–3070 cm^{-1} range, and second in 2960–3000 cm^{-1} range can be assigned to V =C–H (alkenyl) and sp^3 V C–H (alkane) functionalities, respectively. The aromatic C=C band was appeared in the 1175–1200 cm^{-1} range and the amine C–N band was found in the 1420–1460 cm^{-1} range.

^{1}H NMR spectra showed a multiplet in the region δ 2.93–4.09 ppm, in all compounds, except **TC$_{1-5}$**, which are attributable to diastereotopic methylene protons of the pyrrolidine ring. Compounds **TC$_{1-5}$** showed, in fact, a singlet at around δ 2.75 ppm, due to *N*-methyl proton. A characteristic ^{13}C NMR signal of **TC$_{1-20}$** carbonyl carbon was appeared at around δ 173 ppm. A doublet appeared in the δ 4.00–4.80 ppm range, with *J* value in the 7.6–8.4 Hz range, can be attributed well to a bridge-head proton **10b** or **12b,** suggesting that there exists a *cis*-fusion between central two pyrrolidine rings, in all the compounds. The bridge-head proton **2** or **4b** is however *trans* with respect to proton **10b** or **12b,** appeared at δ 4.00–4.40 ppm, except in **TC$_{22}$ $_{\&}$ $_{24}$**. In **TC$_{22}$ $_{\&}$ $_{24}$** it is *cis* oriented.

The stereochemical aspect of the reaction was also studied by taking 2D NMR experiments; DQF–COSY and NOESY of representative compound **TC$_6$.** The COSY and NOESY also helped determine the geometry of the compound. **Fig 1** shows excellent correlation between different kinds of protons in **TC₆**. Thus, the *cis*-fusion between two pyrrolidine rings can be confirmed by 2D NMR experiments.

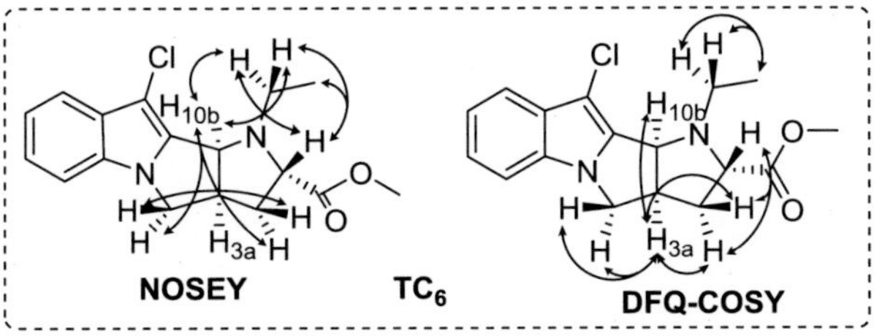

Fig 1 Characteristic COSY and NOESY spectra, **TC₆**.

From the NOESY spectrum, proton H_{10b} is found in the close vicinity of the proton H_{3a} in compound **TC₆**, it supported the *cis*-fusion between two pyrrolidine rings, further. One of the N-methylene protons correlates to H_2 and H_{10b}, respectively, and N-methyl proton to H_2. DQF–COSY experiment supported the same.

Molecular weights and proposed structures of heterocycles were also confirmed based on mass spectrums. ESI/MS spectra of compounds **TC₂**, **TC₉**, **TC₁₈** and **TC₂₀** gave molecular ion peaks 319.10[M]⁺, 346.00 [M]⁺, 430.90 [M]⁺ and 446.10 [M]⁺ (m/z values), respectively, as expected from their corresponding molecular formulas $C_{17}H_{19}ClN_2O_2$, $C_{19}H_{23}ClN_2O_2$, $C_{23}H_{30}ClN_3O_3$ and $C_{24}H_{32}ClN_3O_3$.

4.3   Single crystal X–ray data.[2]

The stereochemistry of all heterocycles was finally confirmed by the single crystal X–ray diffraction analysis. Single crystals of compounds **TC₁₂** and **TC₂₄** were developed in re-crystallizing solvent EA following a slow evaporation method.

4.3.1 The single crystal X–ray data of compounds TC$_{12}$ and TC$_{24}$.

The *ORTEP* views of compounds **TC$_{12}$** and **TC$_{24}$** are shown in **Fig 2** and **3** respectively.

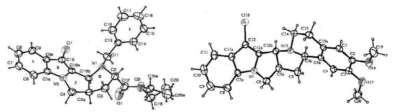

Triclinic, P-1 (CCDC No. **960921**)   Monoclinic, P 2$_{1/n}$ (CCDC No. **1018001**)

Fig 2 ORTEP view of the compound **TC$_{12}$**.   **Fig 3** ORTEP view of the compound **TC$_{24}$**.

4.4 Biological evaluation

All heterocycles were tested for their *in vitro* antimicrobial, antituberculosis and antioxidant activities, and standard drugs listed below were used to compare the screening results.

➢ **Reference drugs used for antimicrobial studies (MIC, μgmL^{-1})**

Drugs	Gram positive bacteria			Gram negative bacteria		Fungi		
	B.S. MTCC 441	C.T. MTCC 449	S.P. MTCC 1936	E.C. MTCC 443	S.T. MTCC 98	V.C. MTCC 3906	A.F. MTCC 3008	C.A MTCC 227
Ampicillin	250	250	100	100	100	100	–	–
Chloramphenicol	50	50	50	50	50	50	–	–
Ciprofloxacin	50	100	50	25	25	25	–	–
Norfloxacin	100	50	10	10	10	10	–	–
Nystatin	–	–	–	–	–	–	100	100
Griseofulvin	–	–	–	–	–	–	100	500

B.S.: *Bacillus subtilis*, C.T.: *Clostridium tetani*, S.P.: *Streptococcus pneumoniae*,

E.C.: *Escherichia coli*, S.T.: *Salmonella typhi*, V.C.: *Vibrio cholerae*,

A.F.: *Aspergillus fumigatus*, C.A.: *Candida albicans*

MTCC: Microbial Type Culture Collection

MIC: Minimum Inhibition Concentration

"–" represent drugs not tested.

➢ **Reference drugs used for antituberculosis studies**

Drugs	% Inhibition	Actual MIC (µg/mL)
Isoniazide	99	0.2
Rifampicin	98	40

➢ **Drugs used for antioxidant studies**

Ascorbic acid was used as a standard to determine total antioxidant power of the compound. Some 200 μL, 1mmol solution of ascorbic acid was tested. Optical density was 2.488 at 593 nm, and used to calculate the FRAP value.

Table 1 displays *in vitro* **antimicrobial** screening test results obtained in case of all the compounds. It showed majority of compounds are good to moderate in their resistivity against bacteria, at least, in comparison with one of the standard reference drugs **Ampicillin**. Some compounds with activity more than ampicillin or comparable with other potent reference drugs were also found in the present study. Analyzing results with a question that maximum how many bacterium species a compound resist against effectively, it was found that a compound covered a panel of maximum five species comparing and considering the potency of at least **Ampicillin**. Examples include **TC** $_{3, 13 \& 20}$, **TC** $_{23}$. Among them, **TC$_3$** revealed excellent activity against Gram-positive *Clostridium tetani* (reaching to the potency of even more potent **Norfloxacin**) and Gram-negative *Escherichia coli* bacteria (reaching to more potent **Chloramphenicol** in activity). Similarly, compound **TC$_{13}$** resembled more potent **Norfloxacin** drug in activity, against Gram-positive *Bacillus subtilis* bacteria. Compounds **TC$_{1, 4 \& 6}$ TC$_{8-9, 15-18 \& 21}$**, on the other hand, had a better resistance against maximum four types of bacteria. Among them, **TC$_{1, 4 \& 8}$, TC$_{16}$** were found highly active against Gram +Ve *Clostridium tetani* bacteria, with MIC values on line with standard drug **Ciprofloxacin** which is more potent than **Ampicillin**. Compounds **TC$_4$** and **TC$_{16}$** with **Chloramphenicol**–equivalent potency was found active against *Bacillus subtilis* and *Salmonella typhi* bacteria respectively. It is noted that both **Ciprofloxacin** and compound **TC$_4$** have similar MIC value against *Bacillus subtilis* bacteria. Those having activity against maximum three bacterium species include **TC$_{2, 5 \& 7}$, TC$_{10, 12 \& 14}$, TC$_9$**. Among them **TC$_7$, TC$_{10}$** and **TC$_{14}$** are very close to standard **ciprofloxacin** in potency, against *Clostridium tetani* bacteria. A few of the compounds were also good in antifungal activity, particularly, against *Candida albicans* fungus. Examples include **TC$_{1, 5 \& 8}$, TC$_{10, 13 \& 19}$** all having **griseofulvin**-equivalent power. As anti-fungal agents, compounds **TC$_4$, TC$_{16}$** and **TC$_{17}$** are relatively more active.

Anti-tubercular activity of compounds **TC$_{14}$** and **TC$_{17}$** was found highest against *M. tuberculosis H37Rv* bacterial species, for that growth inhibition values were observed in the 90-100% range. Compounds **TC$_{6, 11 \& 13}$, TC$_{16 \& 24}$** showed growth inhibition in the 80-90% range.

FRAP values of majority of the heterocycles are around 225 (mmol/100 gm), indicating that they have good **anti-oxidant** power. However, compound **TC$_{17}$** revealed a remarkable activity.

The **antiproliferative activity** of **TC$_{1-24}$** was determined against a panel of six representative human tumor cell lines; A549 (lung), HeLa (cervix), SW1573 (lung), T-47D (breast) and WiDr (colon), using the SRB assay.[28] The experimental GI$_{50}$ values are given in **Table 3** and compared with those of standard drugs **Cisplatin, Etoposide** and **Camptothecin** after 48 h of treatment. Taken as whole, pyrrolo-fused-indoles **TC$_{1-10}$** were the most active in series of compounds, with activity against all cell lines tested. In this particular context, compounds **TC$_{1-10}$** showed excellent resistivity against HeLa cells, with GI$_{50}$ values in the range **3.1–14** μM and comparable to those of **Cisplatin** (2.0 μM) or **Etoposide** (3.3 μM). No significant differences were observed between methyl and ethyl ester derivatives. In contrast, the presence of a benzyl group or a morpholine substituent on the pyrrol nitrogen produced a severe loss in the activity. Among the pyrrolo-fused-isoquinolines, cycloadduct **TC$_{23}$** revealed best antiproliferative activity results, against all the cell lines with GI$_{50}$ values in the **9.5– 18** μM range. This is a relevant result, since the class of adduct correlates to selectivity towards cancer cell lines.

Analysing N-fused indoles derived from amino acid esters, structurally, it reveals that methyl at pyrolidine nitrogen confers heterocycles with enhanced resistivity against Clostridium tetani and Escherichia coli bacteria, when carbpropoxy moiety is present at carbon next to this nitrogen. Pyrrolidine with morpholine or ethyl moieties at nitrogen in combination with carbmethoxy group also had a similar effect against these bacteria. Carbbutoxy moiety, on the other hand, had very less effect on activity, irrespective of substituent present at pyrrolidine nitrogen. Antiproliferative activity, nevertheless, seemed to be altered very less taking any ester component with N-methyl/N-ethyl pyrrolidine ring, against HeLa (cervix) cell lines. In addition, N-ethyl pyrrolidine with any ester component had similar effect against Widr(colon) cell lines. The N-fused indoles derived from tetrahydreoisoquinolines showed remarkable bioactivities as well. In general, heterocycles derived from electron releasing methoxy substituted-tetrahyderoisoquinoline are excellent in antibacterial, antitubercular, antioxidant and antiproliferative activities, compared to the ones derived from simple tetrahydro isoquinoline.

Table 1 Antimicrobial activity screening test results of pyrrolo-fused-indoles and isoquinolines **TC$_{1-24}$**.

Compounds	Substitutions			Antimicrobial Activity (MIC μgmL^{-1})							
				Gram positive bacteria			Gram negative bacteria			Fungi	
	R	R$_1$	R$_2$	S.p.	C.t.	B.s.	S.t.	V.c.	E.c.	A.f.	C.a.
TC$_1$	Me	Me	—	250	**100**	250	250	**125**	**100**	>500	500
TC$_2$	Me	Et	—	200	**125**	**100**	500	**100**	200	>500	>500
TC$_3$	Me	nPr	—	250	**62.5**	200	125	125	**62.5**	>500	>500
TC$_4$	Me	iPr	—	**100**	**100**	**62.5**	200	200	**100**	>500	250
TC$_5$	Me	nBu	—	**100**	250	**100**	250	250	250	>500	500
TC$_6$	Et	Me	—	**125**	200	200	200	250	**62.5**	>500	>500
TC$_7$	Et	Et	—	**125**	**100**	250	200	200	200	>500	>500
TC$_8$	Et	nPr	—	**100**	**100**	250	250	250	**125**	>500	500
TC$_9$	Et	iPr	—	**125**	200	250	250	250	**125**	>500	>500
TC$_{10}$	Et	nBu	—	200	**125**	200	200	250	**100**	>500	500
TC$_{11}$	Bn	Me	—	200	**200**	**125**	250	250	200	>500	>500
TC$_{12}$	Bn	Et	—	250	**250**	200	250	200	**100**	>500	>500
TC$_{13}$	Bn	nPr	—	**100**	250	**100**	100	125	200	>500	500

	R	R₁	R₂								
TC₁₄	Bn	iPr	—	250	**100**	200	**100**	200	250	>500	>500
TC₁₅	Bn	nBu	—	200	**250**	100	**100**	125	250	>500	>500
TC₁₆	Mp	Me	—	**125**	**125**	100	**62.5**	200	200	500	250
TC₁₇	Mp	Et	—	200	**200**	100	125	125	250	500	250
TC₁₈	Mp	nPr	—	500	**250**	250	**125**	250	100	>500	>500
TC₁₉	Mp	iPr	—	200	**250**	100	125	250	250	>1000	500
TC₂₀	Mp	nBu	—	500	**200**	250	**125**	125	100	500	>500
TC₂₁	—	—	H	**100**	200	250	**100**	200	250	500	500
TC₂₂	—	—	H	500	**250**	100	200	250	200	250	500
TC₂₃	—	—	OMe	200	**125**	125	125	100	100	>500	>500
TC₂₄	—	—	OMe	250	**200**	200	200	250	250	500	>500
Ampicillin				250	250	100	100	100	100	--	--
Chloramphenicol				50	50	50	50	50	50	--	--
Ciprofloxacin				50	100	50	25	25	25	--	--
Norfloxacin				100	50	10	10	10	10	--	--
Nystatin				--	--	--	--	--	--	100	100
Griseofulvin				--	--	--	--	--	--	100	500

Table 2   Antituberculosis and antioxidant activity screening test results of pyrrolo-fused-indoles and isoquinolines TC₁₋₂₄.

Compounds	Substitutions			Anti TB	Antioxidant
				% Inhibition (MIC μgmL^{-1})	mM/100g
	R	R₁	R₂		
TC₁	Me	Me	—	46%	225.11
TC₂	Me	Et	—	20%	220.49
TC₃	Me	nPr	—	13%	215.27
TC₄	Me	iPr	—	47%	216.87
TC₅	Me	nBu	—	57%	227.32
TC₆	Et	Me	—	**84%**	219.28
TC₇	Et	Et	—	59%	217.28
TC₈	Et	nPr	—	58%	210.25
TC₉	Et	iPr	—	33%	213.26
TC₁₀	Et	nBu	—	47%	222.90
TC₁₁	Bn	Me	—	80%	213.46
TC₁₂	Bn	Et	—	33%	228.72
TC₁₃	Bn	nPr	—	**87%**	234.95

TC$_{14}$	Bn	^{i}Pr	—	**91%**	236.35
TC$_{15}$	Bn	^{n}Bu	—	12%	245.99
TC$_{16}$	Mp	Me	—	**84%**	232.14
TC$_{17}$	Mp	Et	—	**92%**	**285.15**
TC$_{18}$	Mp	^{n}Pr	—	25%	236.15
TC$_{19}$	Mp	^{i}Pr	—	74%	239.37
TC$_{20}$	Mp	^{n}Bu	—	65%	241.37
TC$_{21}$	—	—	H	65%	224.51
TC$_{22}$	—	—	H	78%	252.22
TC$_{23}$	—	—	OMe	58%	254.23
TC$_{24}$	—	—	OMe	**88%**	217.28
Isoniazide	—	—	—	**99%**	--

Table 3   Antiproliferative activity (GI$_{50}$) against human solid tumor cells

Compounds	Antiproliferative activity (GI$_{50}$)				
	Cell line *(origin)*				
	A549 *(lung)*	**HeLa** *(cervix)*	**SW1573** *(lung)*	**T-47D** *(breast)*	**WiDr** *(colon)*
TC$_1$	>100	**3.3** (±0.4)	68 (±45)	67 (±47)	66 (±48)
TC$_2$	61 (±18)	**5.7** (±1.6)	92 (±11)	84 (±29)	76 (±37)
TC$_3$	>100	**3.9** (±1.2)	53 (±26)	66 (±49)	62 (±54)
TC$_4$	41	**8.3** (±3.2)	29 (±0.2)	**26** (±6.5)	**22** (±4.0)
TC$_5$	32 (±8.8)	**9.9** (±5.2)	32 (±2.3)	**25** (±5.8)	**23** (±4.1)
TC$_6$	33 (±3.0)	**14** (±1.4)	25 (±3.3)	**27** (±4.8)	**24** (±1.7)
TC$_7$	43 (±19)	**3.4** (±0.7)	33 (±4.6)	29 (±16.0)	**28** (±10)
TC$_8$	47 (±19)	**3.6** (±0.3)	41 (±4.3)	37 (±13.0)	35 (±7.8)
TC$_9$	28 (±6.9)	13 (±2.8)	27 (±3.5)	**25** (±9.4)	**27** (±4.3)
TC$_{10}$	40 (±20)	**3.1** (±0.4)	32 (±2.5)	**26** (±4.7)	**24** (±8.9)
TC$_{13}$	>100	41 (±30)	>100	54 (±21)	>100
TC$_{14}$	>100	48 (±36)	>100	57 (±43)	89 (±15)
TC$_{15}$	>100	>100	>100	>100	>100
TC$_{16}$	>100	>100	>100	>100	>100
TC$_{21}$	89 (±16)	33 (±6.8)	>100	53 (±14)	73 (±39)
TC$_{22}$	>100	84 (±21)	>100	>100	>100
TC$_{23}$	18 (±14)	**9.5** (±7.1)	**17** (±1.9)	**15** (±7.3)	**14** (±2.4)
Cisplatin	-	2.0 (±0.3)	3.0 (±0.4)	15 (±2.3)	26 (±5.3)
Etoposide	-	3.3 (±1.6)	14 (±1.5)	22 (±5.5)	23 (±3.1)
Camptothecin	-	0.6 (±0.4)	0.25 (±0.12)	2.0 ±(0.5)	1.8 (±0.7)

4.5 Characterization

TC$_1$	Methyl-10-chloro-1-methyl-1,2,3,3a,4,10b-hexahydropyrrolo-[2',3':3,4] pyrrolo[1,2-*a*]indole-2-carboxylate	
Molecular Formula	C$_{16}$H$_{17}$ClN$_2$O$_2$	
M.P.	118–120 °C	
Mol wt. (gm/mole)	304.97	
Elemental Analysis	C　　　H　　　N	
Cal	63.05　　5.62　　9.19	
Obs	63.35　　5.27　　9.42	
1H–NMR δ ppm (CDCl$_3$)	2.20(1H, m, 3'–H), 2.41(1H, ddd, J = 13.6, 8.2, 5.6 Hz, 3–H), 2.76(3H, s, –NCH$_3$), 3.68(1H, t, J = 5.8Hz, 2–H), 3.78(4H, m, –OCH$_3$; 1H, 3a–H), 3.97 (1H, ddd, J = 9.8, 4.2, 1.4 Hz, 4–H), 4.19 (1H, m, 4'–H), 4.81(1H, d, J = 7.4 Hz, 10b–H), 7.15–7.22(3H, m, Ar–H), 7.70(1H, dd, J = 7.6, 1.0 Hz, 9–H)	
13C–NMR δ ppm (CDCl$_3$)	36.70, 37.11, 45.34, 50.80, 51.36, 64.83, 67.01, 98.27, 110.03, 118.61, 120.10, 122.22, 130.21, 131.93, 138.58, 172.87	
FT–IR: V_{max} cm^{-1} (KBr)	3058, 2983, 1732, 1630, 1455, 1178, 1039, 731, 600	

TC$_2$	Ethyl-10-chloro-1-methyl-1,2,3,3a,4,10b-hexahydropyrrolo[2',3':3,4] pyrrolo[1,2-*a*]indole-2-carboxylate	
Molecular Formula	C$_{17}$H$_{19}$ClN$_2$O$_2$	
M.P.	102–104°C	
Mol wt. (gm/mole)	319.10	
Elemental Analysis	C　　　H　　　N	
Cal	64.05　　6.01　　8.79	
Obs	64.21　　6.17　　8.62	
1H–NMR δ ppm (CDCl$_3$)	1.33(3H, t, J = 6.4 Hz, –OCH$_3$), 2.21(1H, m, 3'–H), 2.40(1H, ddd, J = 13.2, 8.4, 5.2 Hz, 3–H), 2.75(3H, s, –NCH$_3$), 3.66(1H, t, J = 5.6 Hz, 2–H), 3.79(1H, m, 3a–H), 3.96 (1H, ddd, J = 9.9, 4.0, 1.2 Hz, 4–H), 4.23(3H, m, –OCH$_2$; 4'–H), 4.82(1H, d, J = 7.6 Hz, 10b–H), 7.15–7.22(3H, m, Ar–H), 7.60(1H, dd, J = 7.8, 0.8 Hz, 9–H)	
13C–NMR δ ppm (CDCl$_3$)	14.37, 36.72, 37.10, 45.33, 50.81, 60.67, 64.84, 67.02, 98.27, 110.04, 118.60, 120.11, 122.22, 130.22, 131.92, 138.59, 172.86	
FT–IR: V_{max} cm^{-1} (KBr)	3059, 2983, 1730, 1631, 1455, 1177, 1038, 732, 602	

TC₃	Propyl-10-chloro-1-methyl-1,2,3,3a,4,10b-hexahydropyrrolo-[2',3':3,4] pyrrolo[1,2-a]indole-2-carboxylate		
Molecular Formula	$C_{18}H_{21}ClN_2O_2$		
M.P.	106–108°C		
Mol wt. (gm/mole)	332.92		
Elemental Analysis	C	H	N
Cal	64.96	6.36	8.42
Obs	64.75	6.47	8.62
¹H–NMR δ ppm (CDCl₃)	0.99(3H, t, J = 7.8 Hz, –OCH₃), 1.75(2H, m, –CH₂), 2.19(1H, m, 3'–H), 2.42(1H, ddd, J = 13.8, 8.2, 5.4 Hz, 3–H), 2.74(3H, s, –NCH₃), 3.70(1H, t, J = 5.4 Hz, 2–H), 3.79(1H, m, 3a–H), 3.98(1H, ddd, J = 9.6, 4.0, 1.6 Hz, 4–H), 4.17(3H, m, –OCH₂; 4'–H), 4.83(1H, d, J = 7.8 Hz, 10b–H), 7.16–7.22(3H, m, Ar–H), 7.71(1H, dd, J = 7.6, 1.0 Hz, 9–H)		
¹³C–NMR δ ppm (CDCl₃)	10.43, 22.10, 36.69, 37.09, 45.33, 50.79, 64.81, 66.73, 67.03, 98.25, 110.02, 118.59, 120.11, 122.20, 130.22, 131.94, 138.11, 172.86		
FT–IR: V_{max} cm⁻¹ (KBr)	3059, 2984, 1732, 1630, 1457, 1179, 1039, 730, 599		

TC₄	Isopropyl-10-chloro-1-methyl-1,2,3,3a,4,10b-hexahydropyrrolo-[2',3':3,4]pyrrolo[1,2-a]indole-2-carboxylate		
Molecular Formula	$C_{18}H_{21}ClN_2O_2$		
M.P.	107–109°C		
Mol wt. (gm/mole)	332.94		
Elemental Analysis	C	H	N
Cal	64.96	6.36	8.42
Obs	64.88	6.25	8.34
¹H–NMR δ ppm (CDCl₃)	1.30(3H, d, J = 2.4 Hz, –OCH₃), 1.32(3H, d, J = 2.4 Hz, –OCH₃), 2.20(1H, m, 3'–H), 2.39(1H, ddd, J = 12.8, 8.2, 5.0 Hz, 3–H), 2.74(3H, s, –NCH₃), 3.67(1H, t, J = 5.8Hz, 2–H), 3.80(1H, m, 3a–H), 3.96 (1H, ddd, J = 9.8, 4.0, 1.2 Hz, 4–H), 4.18(1H, dd, J = 16.8, 7.8 Hz, 4'–H), 4.81(1H, d, J = 7.4 Hz, H₁₀b), 5.12(1H, sep, J = 6.4 Hz, –OCH), 7.15–7.21(3H, m, Ar–H), 7.61(1H, dd, J = 7.6, 0.8 Hz, 9–H)		
¹³C–NMR δ ppm (CDCl₃)	21.29, 36.70, 37.07, 45.29, 50.79, 64.81, 67.05, 68.03, 98.28, 110.05, 118.58, 120.13, 122.25, 130.25, 131.94, 138.61, 172.87		
FT–IR: V_{max} cm⁻¹ (KBr)	3058, 2984, 1732, 1632, 1456, 1179, 1038, 732, 599		

TC5	Butyl-10-chloro-1-methyl-1,2,3,3a,4,10b-hexahydropyrrolo[2',3':3,4] pyrrolo[1,2-*a*]indole-2-carboxylate	
Molecular Formula	$C_{19}H_{23}ClN_2O_2$	
M.P.	92–94°C	
Mol wt. (gm/mole)	347.04	
Elemental Analysis	C H N	
Cal	65.79 6.68 8.08	
Obs	65.55 6.57 8.22	
1H–NMR δ ppm (CDCl3)	0.99(3H, t, J = 7.8 Hz, –OCH3), 1.43(2H, sex, J = 7.6 Hz, –CH2), 1.68(2H, quin, J = 7.2 Hz, –CH2), 2.22(1H, m, 3'–H), 2.39(1H, ddd, J = 13.6, 8.4, 5.6 Hz, 3–H), 2.73(3H, s, –NCH3), 3.70(1H, t, J = 5.4 Hz, 2–H), 3.78(1H, m, 3a–H), 3.97 (1H, ddd, J = 9.4, 4.2, 1.8 Hz, 4–H), 4.19 (3H, m, –OCH2; 4'–H), 4.81(1H, d, J = 7.6 Hz, 10b–H), 7.15–7.21(3H, m, Ar–H), 7.70(1H, dd, J = 7.4, 1.2 Hz, 9–H)	
13C–NMR δ ppm (CDCl3)	10.46, 19.24, 30.77, 36.70, 37.10, 45.32, 50.78, 64.79, 66.78, 67.02, 98.23, 110.01, 118.58, 120.12, 122.21, 130.20, 131.93, 138.58, 172.88	
FT–IR: V_{max} cm^{-1} (KBr)	3060, 2982, 1732, 1630, 1456, 1178, 1039, 731, 600	

TC6	Methyl-10-chloro-1-ethyl-1,2,3,3a,4,10b-hexahydropyrrolo[2',3':3,4] pyrrolo[1,2-*a*]indole-2-carboxylate	
Molecular Formula	$C_{17}H_{19}ClN_2O_2$	
M.P.	84–86°C	
Mol wt. (gm/mole)	319.10	
Elemental Analysis	C H N	
Cal	64.05 6.01 8.79	
Obs	63.88 6.23 8.94	
1H–NMR δ ppm (CDCl3)	1.20(3H, t, J = 7.2 Hz, –NCH3), 2.19(1H, m, 3'–H), 2.40(1H, ddd, J = 12.8, 9.4, 3.2 Hz, 3–H), 2.95(1H, m, –NCH2), 3.35(1H, m, –NCH2), 3.78(4H, m, –OCH3; 3a–H), 3.96(2H, m, 2–H; 4–H), 4.18(1H, dd, J = 10, 8.4 Hz, 4'–H), 4.82(1H, d, J = 8Hz, 10b–H), 7.17–7.22(3H, m, Ar–H), 7.60(1H, dd, J = 7.6, 1.6 Hz, 9–H)	
13C–NMR δ ppm (CDCl3)	13.84, 35.95, 44.24, 45.03, 50.51, 51.38, 63.91, 64.14, 97.78, 110.00, 118.57, 120.03, 122.15, 130.19, 131.83, 139.46, 173.83	
FT–IR: V_{max} cm^{-1} (KBr)	3051, 2975, 1728, 1631, 1453, 1197, 1169, 738, 650	

TC₇	Ethyl-10-chloro-1-ethyl-1,2,3,3a,4,10b-hexahydropyrrolo[2',3':3,4] pyrrolo[1,2-a]indole-2-carboxylate

Molecular Formula	$C_{18}H_{21}ClN_2O_2$	
M.P.	92–94°C	
Mol wt. (gm/mole)	332.82	

Elemental Analysis	C	H	N
Cal	64.96	6.36	8.42
Obs	64.86	6.53	8.74

¹H–NMR δ ppm (CDCl₃)	1.22(3H, t, J = 7.2 Hz, –NCH₃), 1.32(3H, t, J = 6.8 Hz, –OCH₃), 2.20(1H, m, 3'–H), 2.42(1H, ddd, J = 12.4, 9.6, 3.4 Hz, 3–H), 2.94(1H, m, –NCH₂), 3.32(1H, m, –NCH₂), 3.79(1H , m, 3a–H), 3.98(2H, m, 2–H; 4–H), 4.20(3H, m, 4'–H; –OCH₂), 4.80(1H, d, J = 8.2 Hz, 10b–H), 7.15–7.21(3H, m, Ar–H), 7.62(1H, dd, J = 7.8, 1.4 Hz, 9–H)
¹³C–NMR δ ppm (CDCl₃)	13.85, 14.38, 35.94, 44.28, 45.06, 50.49, 60.38, 63.93, 64.11, 97.75, 112.02, 118.58, 120.02, 122.14, 130.19, 131.85, 139.48, 173.85
FT–IR: V_{max} cm⁻¹ (KBr)	3053, 2973, 1730, 1630, 1450, 1248, 1198, 1170, 739, 652

TC₈	Propyl-10-chloro-1-ethyl-1,2,3,3a,4,10b-hexahydropyrrolo[2',3':3,4] pyrrolo[1,2-a]indole-2-carboxylate

Molecular Formula	$C_{19}H_{23}ClN_2O_2$	
M.P.	64–66°C,	
Mol wt. (gm/mole)	346.87	

Elemental Analysis	C	H	N
Cal	65.79	6.68	8.08
Obs	65.89	6.43	7.89

¹H–NMR δ ppm (CDCl₃)	0.99(3H, t, J = 7.4 Hz, –OCH₃), 1.21(3H, t, J = 7.4 Hz, –NCH₃), 1.72(2H, m, –CH₂), 2.22(1H, m, 3'–H), 2.41(1H, ddd, J = 12.6, 9.2, 3.4 Hz, 3–H), 2.95(1H, m, –NCH₂), 3.30(1H, m, –NCH₂), 3.78(1H, m, 3a–H), 3.97(2H, m, 2–H; 4–H), 4.18(3H, m, 4'–H; –OCH₂), 4.79(1H, d, J = 7.8 Hz, 10b–H), 7.21(3H, m, Ar–H), 7.62(1H, dd, J = 7.4, 1.6 Hz, 9–H)
¹³C–NMR δ ppm (CDCl₃)	10.44, 13.82, 22.12, 35.95, 44.26, 45.09, 50.52, 63.91, 64.15, 66.38, 97.72, 112.02, 118.55, 120.05, 122.13, 130.16, 131.81, 139.42, 173.89
FT–IR: V_{max} cm⁻¹ (KBr)	3051, 2975, 1728, 1631, 1453, 1197, 1169, 738, 650

TC9	Isopropyl-10-chloro-1-ethyl-1,2,3,3a,4,10b-hexahydropyrrolo-[2',3':3,4] pyrrolo[1,2-a]indole-2-carboxylate	
Molecular Formula	$C_{19}H_{23}ClN_2O_2$	
M.P.	62–63°C	
Mol wt. (gm/mole)	346.00	
Elemental Analysis	C H N	
Cal	65.79 6.68 8.08	
Obs	65.88 6.73 8.16	
¹H–NMR δ ppm (CDCl₃)	1.21(3H, t, J = 7.2 Hz, –NCH₃), 1.31(3H, d, J = 2.4 Hz, –OCH₃), 1.32(3H, d, J = 2.4 Hz, –OCH₃), 2.19(1H, m, 3'–H), 2.39(1H, ddd, J = 13.4, 9.2, 2.8 Hz, 3–H), 2.93(1H, m, –NCH₂), 3.37(1H, m, –NCH₂), 3.81(1H, m, 3a–H), 3.93(1H, dd, J = 7.2, 2.8 Hz, 4–H), 3.98(1H, dd, J = 10.2, 4.0 Hz, 2–H), 4.20(1H, dd, J = 17.2, 8.0 Hz, 4'–H), 4.82(1H, d, J = 8.4 Hz, 10b–H), 5.12(1H, sep, J = 6.4 Hz, –OCH), 7.19(3H, m, Ar–H), 7.59(1H, dd, J = 7.8, 1.2 Hz, 9–H).	
¹³C–NMR δ ppm (CDCl₃)	13.52, 21.28, 35.92, 44.65, 45.72, 50.74, 64.16, 64.21, 68.02, 97.49, 110.08, 118.16, 120.18, 122.23, 130.15, 131.65, 139.64, 173.48	
FT–IR: Vmax cm⁻¹ (KBr)	3054, 2967, 1733, 1631, 1456, 1373, 1185, 1107, 737, 656	

TC10	Butyl-10-chloro-1-ethyl-1,2,3,3a,4,10b-hexahydropyrrolo[2',3':3,4] pyrrolo[1,2-a]indole-2-carboxylate	
Molecular Formula	$C_{20}H_{25}ClN_2O_2$	
M.P.	60–62°C	
Mol wt. (gm/mole)	360.88	
Elemental Analysis	C H N	
Cal	66.56 6.98 7.76	
Obs	66.78 6.54 7.98	
¹H–NMR δ ppm (CDCl₃)	0.98(3H, t, J = 7.6 Hz, –OCH₃), 1.19(3H, t, J = 7.8 Hz, –NCH₃), 1.42(2H, sex, J = 7.2 Hz, –CH₂), 1.68(2H, quin, J = 7.2 Hz, –CH₂), 2.21(1H, m, 3'–H), 2.41(1H, ddd, J = 12.8, 9.6, 3.2 Hz, 3–H), 2.94(1H, m, –NCH₂), 3.32(1H, m, –NCH₂), 3.79(1H, m, 3a–H), 3.96(2H, m, 2-H; 4–H), 4.19(3H, m, 4'–H; –OCH₂), 4.80(1H, d, J = 7.6 Hz, 10b–H), 7.15–7.21(3H, m, Ar–H), 7.61(1H, dd, J = 7.2, 1.8 Hz, 9–H)	
¹³C–NMR δ ppm (CDCl₃)	10.42, 13.81, 19.24, 30.76, 35.96, 44.24, 45.07, 50.54, 63.90, 64.13, 66.81, 97.70, 112.03, 118.54, 120.05, 122.11, 130.15, 131.80, 139.41, 173.88	
FT–IR: Vmax cm⁻¹ (KBr)	3052, 2976, 1729, 1630, 1451, 1198, 1170, 1130, 740, 648	

TC$_{11}$	Methyl-1-benzyl-10-chloro-1,2,3,3a,4,10b-hexahydropyrrolo-[2',3':3,4] pyrrolo[1,2-*a*]indole-2-carboxylate	
Molecular Formula	C$_{22}$H$_{21}$ClN$_2$O$_2$	
M.P.	154–156°C	
Mol wt. (gm/mole)	380.87	

Elemental Analysis	C	H	N
Cal	69.38	5.56	7.36
Obs	69.54	5.68	7.42

^{1}H–NMR δppm (CDCl$_3$)	2.16(1H, m, 3'–H), 2.35(1H, ddd, J = 12.4, 9.0, 3.2 Hz, 3–H), 3.67(1H, t, J = 6.2 Hz, 2–H), 3.77(3H, s, –OCH$_3$), 3.84(1H, m, 3a–H), 4.06(2H, m, –NCH$_2$), 4.21(1H, m, 4–H), 4.79(1H, d, J = 13.4 Hz, 4'–H), 5.01(1H, d, J = 8.2 Hz, 10b–H), 7.18–7.31(8H, m, Ar–H), 7.62(1H, d, J = 7.4 Hz, 9–H)
^{13}C–NMR δppm (CDCl$_3$)	35.86, 45.42, 50.18, 51.36, 54.08, 63.94, 64.40, 97.27, 110.03, 118.61, 120.09, 122.12, 127.11, 128.23, 129.02, 130.21, 131.72, 138.86, 140.32, 173.44
FT–IR: V_{max} cm^{-1} (KBr)	3055, 2980, 1723, 1581, 1455, 1369, 1195, 1028, 739, 705, 598

TC$_{12}$	Ethyl1-benzyl-10-chloro-1,2,3,3a,4,10b-hexahydropyrrolo[2',3':3,4] pyrrolo[1,2-*a*]indole-2-carboxylate	
Molecular Formula	C$_{23}$H$_{23}$ClN$_2$O$_2$	
M.P.	124–125°C	
Mol wt. (gm/mole)	395.30	

Elemental Analysis	C	H	N
Cal	69.95	5.87	7.09
Obs	69.74	5.68	7.23

^{1}H–NMR δppm (CDCl$_3$)	1.30(3H, t, J = 7.2 Hz, –OCH$_3$), 2.17(1H, m, 3'–H), 2.34(1H, ddd, J = 12.6, 9.2, 3.0 Hz, 3–H), 3.68(1H, t, J = 6.0 Hz, 2–H), 3.85(1H, m, 3a–H), 4.05(2H, m, –NCH$_2$), 4.20(3H, m, –OCH$_2$; 4–H), 4.68(1H, d, J = 13.6 Hz, 4'–H), 5.01(1H, d, J = 8.0 Hz, 10b–H), 7.18–7.31(8H, m, Ar–H), 7.63(1H, d, J = 7.6 Hz, 9–H)
^{13}C–NMR δppm (CDCl$_3$)	14.41, 35.88, 45.44, 50.19, 54.06, 60.36, 63.93, 64.42, 97.26, 110.05, 118.63, 120.08, 122.13, 127.13, 128.22, 129.00, 130.20, 131.70, 138.87, 140.31, 173.46
FT–IR: V_{max} cm^{-1} (KBr)	3053, 2982, 1720, 1582, 1455, 1371, 1194, 1027, 739, 700, 599

TC₁₃	Propyl-1-benzyl-10-chloro-1,2,3,3a,4,10b-hexahydropyrrolo-[2',3':3,4] pyrrolo[1,2-*a*]indole-2-carboxylate		
Molecular Formula	$C_{24}H_{25}ClN_2O_2$		
M.P.	126–128°C		
Mol wt. (gm/mole)	408.92		
Elemental Analysis	C	H	N
Cal	70.49	6.16	6.85
Obs	70.54	6.38	6.72
¹H–NMR δ ppm (CDCl₃)	1.02(3H, t, J = 7.4Hz, –OCH₃), 1.76(2H, m, –CH₂), 2.18(1H, m, 3'–H), 2.33(1H, ddd, J = 12.8, 9.0, 3.0 Hz, 3–H), 3.69(1H, t, J = 5.8 Hz, 2–H), 3.86(1H, m, 3a–H), 4.09(4H, m, –NCH₂; –OCH₂), 4.19(1H, m, 4–H), 4.81(1H, d, J = 13.6 Hz, 4'–H), 5.01(1H, d, J = 7.8 Hz, 10b–H), 7.19–7.30(8H, m, Ar–H), 7.64(1H, d, J = 7.2 Hz, 9–H)		
¹³C–NMR δ ppm (CDCl₃)	10.45, 22.10, 35.89, 45.45, 50.18, 54.07, 63.92, 64.41, 66.70, 97.25, 110.04, 118.62, 120.07, 122.14, 127.12, 128.21, 129.01, 130.19, 131.71, 138.88, 140.31, 173.45		
FT–IR: V_{max} cm⁻¹ (KBr)	3054, 2981, 1725, 1582, 1457, 1369, 1196, 1029, 738, 702, 601		

TC₁₄	Isopropyl-1-benzyl-10-chloro-1,2,3,3a,4,10b–hexahydropyrrolo-[2',3':3,4]pyrrolo[1,2-*a*]indole-2-carboxylate		
Molecular Formula	$C_{24}H_{25}ClN_2O_2$		
M.P.	132–134°C		
Mol wt. (gm/mole)	408.90		
Elemental Analysis	C	H	N
Cal	70.49	6.16	6.85
Obs	70.54	6.28	6.73
¹H–NMR δ ppm (CDCl₃)	1.31(3H, d, J = 2.4 Hz, –OCH₃), 1.33(3H, d, J = 2.4 Hz, –OCH₃), 2.18(1H, m, 3'–H), 2.36(1H, ddd, J = 12.8, 9.2, 3.2 Hz, 3–H), 3.69(1H, t, J = 6.0 Hz, 2–H), 3.84(1H, m, 3a–H), 4.03(2H, m, –NCH₂), 4.20(1H, m, 4–H), 4.82(1H,d, J = 13.4 Hz, 4'–H), 5.02(1H, d, J = 8.4 Hz, 10b–H), 5.14(1H, sep, J = 6.4 Hz, –OCH), 7.19–7.29(8H, m, Ar–H), 7.65(1H, d, J = 7.8 Hz, 9–H)		
¹³C–NMR δ ppm (CDCl₃)	21.30, 35.85, 45.42, 50.21, 54.05, 63.89, 64.39, 68.06, 97.28, 110.08, 118.61, 120.09, 122.11, 127.14, 128.24, 129.02, 130.19, 131.72, 138.85, 140.35, 173.49		
FT–IR: V_{max} cm⁻¹ (KBr)	3055, 2984, 1719, 1579, 1458, 1369, 1192, 1028, 740, 702, 602		

TC$_{15}$	Butyl-1-benzyl-10-chloro-1,2,3,3a,4,10b-hexahydropyrrolo-[2',3':3,4] pyrrolo[1,2-*a*]indole-2-carboxylate
Molecular Formula	$C_{25}H_{27}ClN_2O_2$
M.P.	96–98°C
Mol wt. (gm/mole)	422.95

Elemental Analysis	C	H	N
Cal	70.99	6.43	6.62
Obs	70.64	6.58	6.92

^{1}H–NMR δppm (CDCl$_3$)	0.99(3H, t, J = 7.8 Hz, –OCH$_3$), 1.46(2H, sex, J = 7.6 Hz, –CH$_2$), 1.66(2H, quin, J = 6.8 Hz, –CH$_2$), 2.16(1H, m, 3'–H), 2.34(1H, ddd, J = 12.6, 8.8, 3.2 Hz, 3–H), 3.68(1H, t, J = 6.2 Hz, 2–H), 3.87(1H, m, 3a–H), 4.06(2H, m, –NCH$_2$), 4.17(3H, m, 4–H; –OCH$_2$), 4.79(1H, d, J = 13.4 Hz, 4'–H), 4.99(1H, d, J = 7.6 Hz, 10b–H), 7.17–7.31(8H, m, Ar–H), 7.62(1H, d, J = 7.6 Hz, 9–H)
^{13}C–NMR δppm (CDCl$_3$)	10.44, 19.24, 30.78, 35.87, 45.46, 50.17, 54.05, 63.93, 64.43, 66.78, 97.27, 110.06, 118.62, 120.08, 122.12, 127.13, 128.23, 129.01, 130.21, 131.70, 138.89, 140.30, 173.47
FT–IR: ν_{max} cm^{-1} (KBr)	3052, 2981, 1722, 1584, 1453, 1372, 1194, 1028, 741, 702, 597

TC$_{16}$	Methyl-10-chloro-1-(2-morpholin-4-ylethyl)1,2,3,3a,4,10b-hexahydro-pyrrolo[2',3':3,4]pyrrolo[1,2-*a*]indole-2-carboxylate
Molecular Formula	$C_{21}H_{26}ClN_3O_3$
M.P.	–
Mol wt. (gm/mole)	404.02

Elemental Analysis	C	H	N
Cal	62.45	6.49	10.40
Obs	62.64	6.38	10.02

^{1}H–NMR δppm (CDCl$_3$)	2.24(1H, m, 3'–H), 2.40–2.62(7H, m, 3–H; –CH$_2$N(CH$_2$)$_2$–), 3.12(1H, m, 11–H), 3.52(1H, m , 11–H), 3.70(5H, m, 3a–H; –CH$_2$OCH$_2$–), 3.79(3H, s, –OCH$_3$), 3.96(1H, dd, J = 10.2, 4.0 Hz, 2–H), 4.07(2H, m, 4–H; 4'–H), 4.86(1H, d, J = 8.4 Hz, 10b–H), 7.17–7.22(3H, m, Ar–H), 7.59(1H, d, J = 7.6 Hz, 9–H)
^{13}C–NMR δppm (CDCl$_3$)	36.15, 44.81, 46.46, 50.35, 51.38, 53.76, 57.74, 64.69, 64.74, 66.26, 97.34, 110.01, 118.46, 120.05, 122.18, 130.04, 131.71, 139.74, 173.59
FT–IR: ν_{max} cm^{-1} (KBr)	3060, 2956, 1727, 1684, 1456, 1338, 1158, 1029, 745, 702, 664, 608

TC₁₇	Ethyl-10-chloro-1-(2-morpholin-4-ylethyl)-1,2,3,3a,4,10b-hexahydro-pyrrolo[2',3':3,4]pyrrolo[1,2-*a*]indole-2-carboxylate

Molecular Formula	$C_{22}H_{28}ClN_3O_3$	
M.P.	—	
Mol wt. (gm/mole)	417.90	

Elemental Analysis	C	H	N
Cal	63.22	6.75	10.05
Obs	63.54	6.58	10.22

¹H–NMR δ ppm (CDCl₃)	1.32(3H, t, J = 7.6 Hz, –OCH₃), 2.23(1H, m, 3'–H), 2.38–2.63(7H, m, 3–H; –CH₂N(CH₂)₂–), 3.13(1H, m, 11–H), 3.51(1H, m, 11–H), 3.71(5H, m, 3a–H; –CH₂OCH₂–), 3.97(1H, dd, J = 10, 4.2 Hz, 2–H), 4.11(4H, m, –OCH₂; 4–H; 4'–H), 4.85(1H, d, J = 8.2 Hz, 10b–H), 7.16–7.22(3H, m, Ar–H), 7.58(1H, d, J = 7.4 Hz, 9–H)
¹³C–NMR δ ppm (CDCl₃)	14.40, 36.14, 44.82, 46.44, 50.36, 53.74, 57.72, 60.56, 64.67, 64.76, 66.25, 97.33, 110.01, 118.47, 120.04, 122.19, 130.04, 131.72, 139.73, 173.60
FT–IR: V_{max} cm⁻¹ (KBr)	3059, 2954, 1729, 1685, 1454, 1336, 1151, 1028, 742, 700, 662, 612

TC₁₈	Propyl-10-chloro-1-(2-morpholin-4-ylethyl)-1,2,3,3a,4,10b-hexahydro-pyrrolo[2',3':3,4]pyrrolo[1,2-*a*]indole-2-carboxylate

Molecular Formula	$C_{23}H_{30}ClN_3O_3$	
M.P.	—	
Mol wt. (gm/mole)	430.90	

Elemental Analysis	C	H	N
Cal	63.95	7.00	9.73
Obs	63.74	6.78	9.96

¹H–NMR δ ppm (CDCl₃)	1.01(3H, t, J = 7.6 Hz, –OCH₃), 1.74(2H, m, –CH₂), 2.25(1H, m, 3'–H), 2.40–2.65(7H, m, 3–H; –CH₂N(CH₂)₂–), 3.12(1H, m, 11–H), 3.50(1H, m, 11–H), 3.70(5H, m, 3a–H; –CH₂OCH₂–), 3.97(1H, dd, J = 10, 4.4 Hz, 2–H), 4.15(4H, m, –OCH₂; 4–H; 4'–H), 4.84(1H, d, J = 8.0 Hz, 10b–H), 7.15–7.21(3H, m, Ar–H), 7.58(1H, d, J = 7.6 Hz, 9–H)
¹³C–NMR δ ppm (CDCl₃)	10.44, 22.11, 36.12, 44.84, 46.24, 50.35, 53.75, 57.73, 64.66, 64.86, 66.15, 66.72, 97.31, 110.00, 118.46, 120.06, 122.21, 130.04, 131.71, 139.74, 173.59
FT–IR: V_{max} cm⁻¹ (KBr)	3059, 2952, 1728, 1632, 1455, 1308, 1116, 1026, 743, 701, 612

TC$_{19}$	Isopropyl-10-chloro-1-(2-morpholin-4-ylethyl)-1,2,3,3a,4,10b-hexa-hydropyrrolo[2',3':3,4]pyrrolo[1,2-a]indole-2-carboxylate
Molecular Formula	C$_{23}$H$_{30}$ClN$_3$O$_3$
M.P.	–
Mol wt. (gm/mole)	430.94

Elemental Analysis	C	H	N
Cal	63.95	7.00	9.73
Obs	63.86	6.81	9.86

^{1}H–NMR δ ppm (CDCl$_3$)	1.32(3H, d, J = 2.4 Hz, –OCH$_3$), 1.34(3H, d, J = 2.4 Hz,- OCH$_3$), 2.24(1H, m, 3'–H), 2.40–2.66(7H, m, 3–H; – CH$_2$N(CH$_2$)$_2$–), 3.12(1H, m, 11–H), 3.51(1H, m, 11–H), 3.71(5H, m, 3a–H; –CH$_2$OCH$_2$–), 3.98(1H, dd, J = 10.2, 4.6 Hz, 2–H), 4.15–4.18(2H, m, 4–H; 4'–H), 4.85(1H, d, J = 8.2 Hz, 10b–H), 5.14(1H, m, –OCH), 7.16–7.23(3H, m, Ar–H), 7.59(1H, d, J = 7.6 Hz, 9–H)
^{13}C–NMR δ ppm (CDCl$_3$)	21.30, 36.13, 44.82, 46.85, 50.34, 53.71, 57.75, 64.67, 64.88, 66.14, 66.17, 97.32, 110.01, 118.46, 120.07, 122.20, 130.05, 131.70, 139.74, 173.62
FT–IR: ν_{max} cm^{-1} (KBr)	3060, 2950, 1729, 1632, 1454, 1309, 1117, 1028, 741, 702, 614

TC$_{20}$	Butyl-10-chloro-1-(2-morpholin-4-ylethyl)-1,2,3,3a,4,10b-hexahydro–pyrrolo[2',3':3,4] pyrrolo[1,2-a]indole-2-carboxylate
Molecular Formula	C$_{24}$H$_{32}$ClN$_3$O$_3$
M.P.	–
Mol wt. (gm/mole)	446.10

Elemental Analysis	C	H	N
Cal	64.63	7.23	9.42
Obs	64.74	7.48	9.76

^{1}H–NMR δ ppm (CDCl$_3$)	0.98(3H, t, J = 8.0 Hz, –OCH$_3$), 1.44(2H, sex, J = 7.6 Hz, – CH$_2$), 1.67(2H, quin, J = 7.2 Hz, –CH$_2$), 2.24(1H, m, 3'–H), 2.50(7H, m, 3–H; –CH$_2$N(CH$_2$)$_2$–), 3.12(1H, m, 11–H), 3.50(1H, m, 11–H), 3.69(5H, m, 3a–H; –CH$_2$OCH$_2$–), 3.98(1H, dd, J = 10.0, 4.0 Hz, 2–H), 4.06(1H, dd, J = 7.6, 2.4 Hz, 4–H), 4.18(3H, m, –OCH$_2$–; 4'–H), 4.84(1H, d, J = 8.4 Hz, 10b–H), 7.17–7.21(3H, m, Ar–H), 7.58(1H, d, J = 7.6 Hz, 9–H)
^{13}C–NMR δ ppm (CDCl$_3$)	10.45, 19.25, 30.77, 36.13, 44.83, 46.32, 50.34, 53.79, 57.79, 64.41, 64.69, 64.98, 66.79, 97.32, 109.99, 118.45, 120.05, 122.20, 130.06, 131.72, 139.76, 173.61
FT–IR: ν_{max} cm^{-1} (KBr)	3060, 2955, 1728, 1685, 1455, 1337, 1152, 1028, 744, 701, 663, 613

TC21	12-Chloro-4b*[R]*,5,5a,12b,14,15-hexahydro-6H-benzo[5',6']pyrrolizino [2',1':4,5]pyrrolo[2,1-*a*]isoquinoline
Molecular Formula	$C_{21}H_{19}ClN_2$
M.P.	118–120°C
Mol wt. (gm/mole)	334.20

Elemental Analysis	C	H	N
Cal	75.33	5.72	8.37
Obs	75.41	5.63	8.42

¹H–NMR δ ppm (CDCl₃)	2.30(1H, m, 5'–H), 2.40(1H, m, 5–H), 2.90(1H, m, 15–H), 3.15(2H, m, 15–H; 14–H), 3.66(2H, m, 14–H; 5a–H), 4.00(1H, t, *J* = 7.6 Hz, 4b–H), 4.10(1H, dd, *J* = 10.2, 3.2 Hz, 6–H), 4.28(1H, t, *J* = 10.0 Hz, 6'–H), 4.93(1H, d, *J* = 7.6 Hz, 12b–H), 7.07–7.26(7H, m, Ar–H), 7.61(1H, d, *J* = 8.0 Hz, 11–H)
¹³C–NMR δ ppm (CDCl₃)	24.96, 34.67, 44.23, 48.40, 51.58, 59.63, 64.40, 99.06, 109.33, 119.31, 121.36, 122.58, 123.27, 125.08, 126.89, 128.24, 128.90, 129.94, 130.85, 137.41, 139.57
FT–IR: *V*max cm⁻¹ (KBr)	3054, 2939, 2817, 1637, 1455, 1321, 1220, 740, 642

TC22	12-Chloro-4b*[S]*,5,5a,12b,14,15-hexahydro-6H-benzo[5',6']pyrrolizino [2',1':4,5]pyrrolo[2,1-*a*]isoquinoline
Molecular Formula	$C_{21}H_{19}ClN_2$
M.P.	152–154°C
Mol wt. (gm/mole)	334.20

Elemental Analysis	C	H	N
Cal	75.33	5.72	8.37
Obs	75.39	5.61	8.45

¹H–NMR δ ppm (CDCl₃)	1.89(1H, m, 5'–H), 2.78(1H, m, 5–H), 2.88(2H, m, 15–H), 3.16(1H, m, 14–H), 3.72(2H, m, 14–H; 5a–H), 3.89(1H, dd, *J* = 10.8, 7.2 Hz, 6–H), 3.97(1H, dd, *J* = 10.0, 6.4 Hz, 6'–H), 4.12(1H, d, *J* = 8.4 Hz, 12b–H), 4.39(1H, t, *J* = 9.6 Hz, 4b–H), 7.08–7.21(7H, m, Ar–H), 7.61(1H, dd, *J* = 7.6, 1.2 Hz, 11–H)
¹³C–NMR δ ppm (CDCl₃)	25.04, 34.61, 44.20, 48.25, 51.58, 59.65, 64.58, 98.96, 109.35, 119.33, 121.36, 122.58, 123.25, 125.22, 127.06, 128.24, 128.93, 130.04, 130.85, 137.45, 139.87
FT–IR: *V*max cm⁻¹ (KBr)	3055, 2938, 2818, 1638, 1454, 1321, 1222, 1138, 742, 640

TC23	12-Chloro-2,3-dimethoxy-4b*[R]*,5,5a,12b,14,15-hexahydro-6H-benzo-[5',6']pyrrolizino[2',1':4,5]pyrrolo[2,1-*a*]isoquinoline
Molecular Formula	$C_{23}H_{23}ClN_2O_2$
M.P.	162–164°C
Mol wt. (gm/mole)	394.54

Elemental Analysis	C	H	N
Cal	69.95	5.87	7.09
Obs	69.81	5.93	7.02

¹H– NMR δ ppm (CDCl₃)	2.32(2H, m, 5–H; 5'–H), 2.80(1H, m, 15–H), 3.13(2H, m, 15–H; 14–H), 3.65(2H, m, 14–H; 5a–H), 3.85(3H, s, –OCH₃), 3.87(3H, s, –OCH₃), 4.01(1H, t, *J* = 7.2 Hz, 4b–H), 4.09(1H, dd, *J* = 10.4, 3.6 Hz, 6–H), 4.27(1H, dd, *J* = 10.2, 8.4 Hz, 6–H·), 4.90(1H, d, *J* = 7.6 Hz, 12b–H), 6.56 (1H, s, 1–H), 6.64 (1H, s, 4–H), 7.15–7.26(3H, m, Ar–H), 7.60(1H, dd, *J* = 7.0, 1.2 Hz, 11–H)
¹³C–NMR δ ppm (CDCl₃)	26.25, 36.88, 40.94, 45.72, 50.63, 55.87, 56.52, 60.49, 64.62, 98.42, 109.72, 111.97, 117.07, 118.60, 121.86, 123.28, 125.21, 128.73, 130.08, 130.31, 132.58, 147.04, 147.57
FT–IR: ν max cm⁻¹ (KBr)	2998, 2934, 2820, 1608, 1516, 1452, 1323, 1219, 1014, 746, 674

TC24	12-Chloro-2,3-dimethoxy-4b*[S]*,5,5a,12b,14,15-hexahydro-6H-benzo-[5',6']-pyrrolizino[2',1':4,5]pyrrolo[2,1-*a*]isoquinoline
Molecular Formula	$C_{23}H_{23}ClN_2O_2$
M.P.	202–204°C
Mol wt. (gm/mole)	394.54

Elemental Analysis	C	H	N
Cal	69.95	5.87	7.09
Obs	69.91	5.90	7.04

¹H–NMR δ ppm (CDCl₃)	1.90(1H, m, 5'–H), 2.79(3H, m, 5–H; 15–H), 3.07(1H, m, 14–H), 3.68(2H, m, 14–H; 5a–H), 3.83(4H, m, –OCH₃; 6–H), 3.88(3H, s, –OCH₃), 3.98 (1H, dd, *J* = 9.6, 6.4 Hz, 6'–H), 4.13(1H, d, *J* = 8.0 Hz, 12b–H), 4.40(1H, t, *J* = 9.6 Hz, 4b–H), 6.58 (1H, s, 1–H), 6.63 (1H, s, 4–H), 7.14–7.20(3H, m, Ar–H), 7.60(1H, d, *J* = 7.6 Hz, 11–H)
¹³C–NMR δ ppm (CDCl₃)	26.20, 36.92, 41.04, 45.78, 50.58, 56.04, 56.64, 60.48, 64.59, 99.02, 110.12, 112.07, 117.09, 118.65, 121.89, 123.35, 125.28, 128.73, 130.08, 130.42, 132.68, 147.24, 147.87
FT–IR: ν max cm⁻¹ (KBr)	3002, 2935, 2818, 1608, 1518, 1454, 1325, 1218, 1012, 748, 672

Mass spectrum of compound TC₂

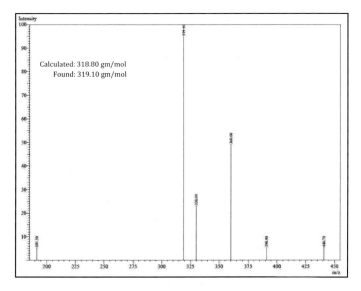

Calculated: 318.80 gm/mol
Found: 319.10 gm/mol

IR spectrum of compound TC₂

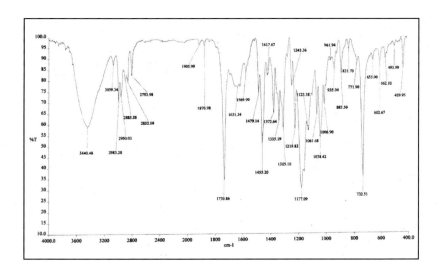

^{1}H NMR spectrum of compound TC$_2$

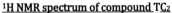

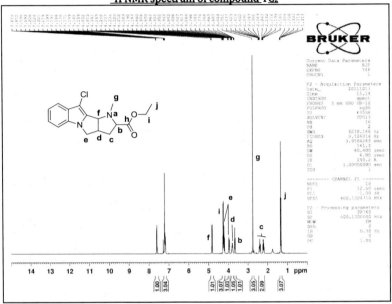

^{13}C NMR spectrum of compound TC$_2$

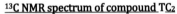

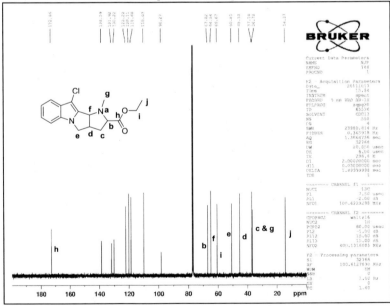

IR spectrum of compound TC₆

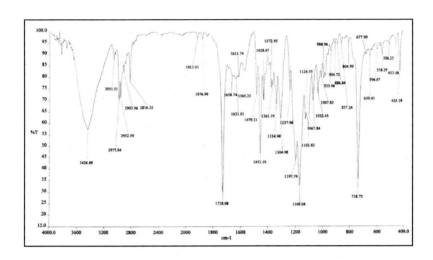

¹H NMR spectrum of compound TC₆

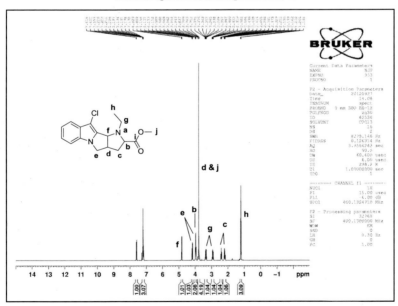

APT spectrum of compound TC₆

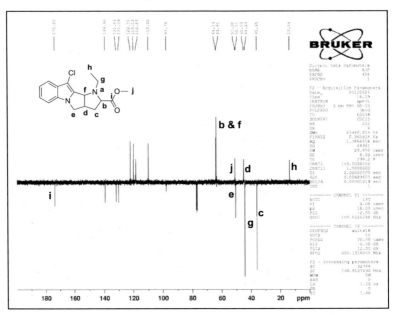

NOESY spectrum of compound TC₆

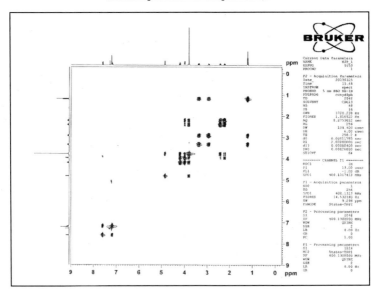

^{1}H NMR spectrum of compound TC$_9$

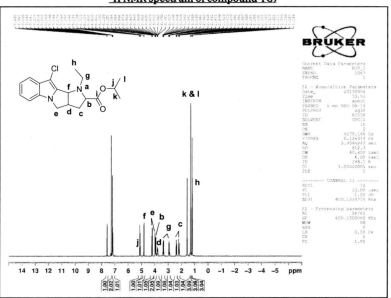

APT spectrum of compound TC$_9$

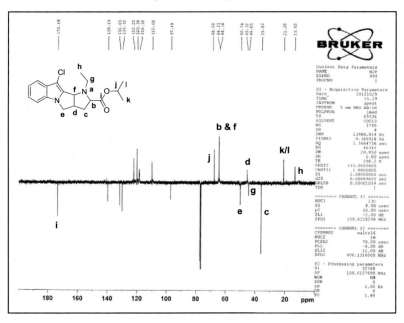

Mass spectrum of compound TC₁₂

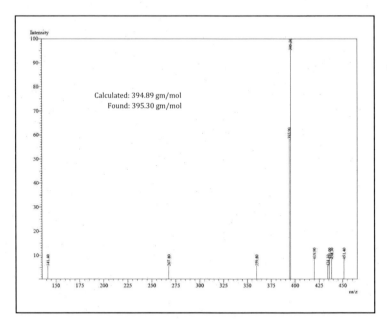

Calculated: 394.89 gm/mol
Found: 395.30 gm/mol

IR spectrum of compound TC₁₂

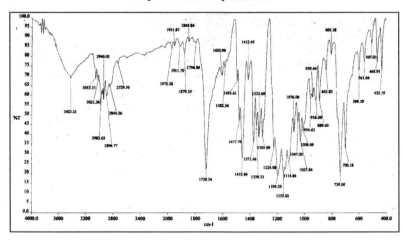

^{1}H NMR spectrum of compound TC$_{12}$

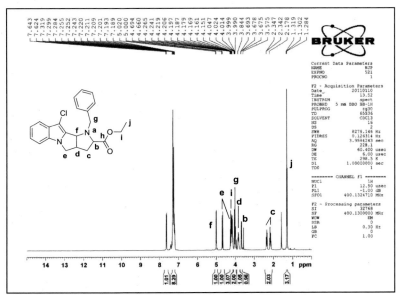

^{13}C NMR spectrum of compound TC$_{12}$

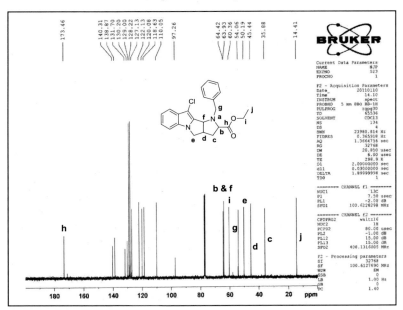

^{1}H NMR spectrum of compound TC$_{18}$

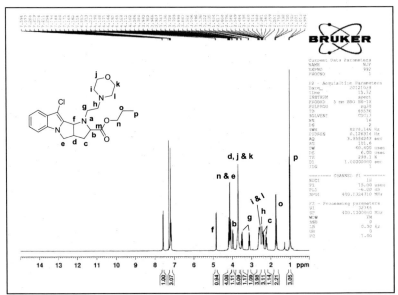

APT spectrum of compound TC$_{18}$

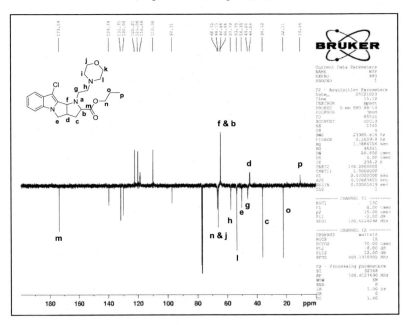

^{1}H NMR spectrum of compound TC$_{20}$

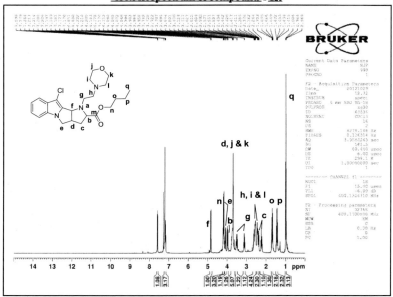

APT spectrum of compound TC$_{20}$

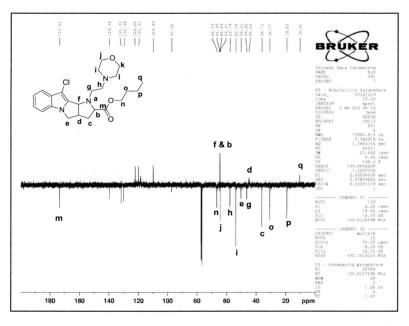

^1H NMR spectrum of compound TC$_{21}$

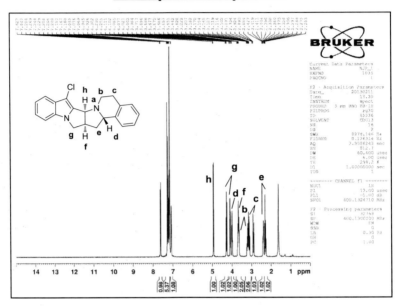

^1H NMR spectrum of compound TC$_{22}$

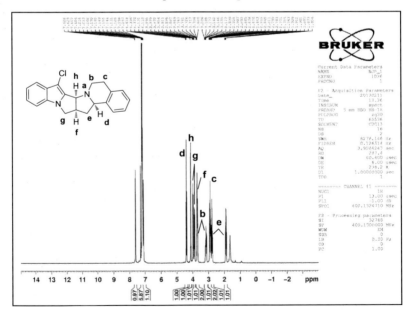

¹H NMR spectrum of compound TC₂₃

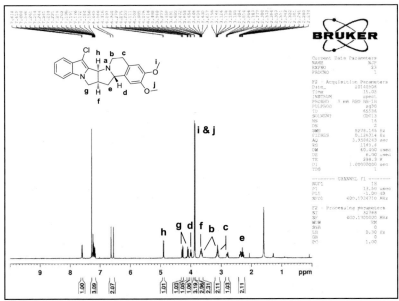

¹H NMR spectrum of compound TC₂₄

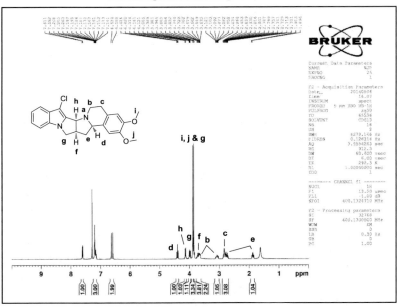

References

1   (a) Kanemasa, S.; Sakamoto K.; Tsuge, O.; *Bull. Chem Soc. Jpn.,* **1989**, *62*, 1960; (b) Khlebnikov, A. F.; Novikov, M. S.; Kostikov R. R.; Kopl, J. *Russ. J. Org. Chem.,* **2005**, *41*, 1341.

2.   (a) Duax, W. L.; Norton, D. A. *Atlas of Steroid Structures*; Plenum : New York, Vol. 1, **1975**; (b) Farrugia, L. J. *J. Appl. Cryst.* **1997**, *30*, 565; (c) Sheldrick, G. M. SHELX97 University of Gottingen, Germany, 1997; (d) Spek, A. L. *Acta. Cryst.* **2009**, *D65*, 148.

Printed by Books on Demand GmbH, Norderstedt / Germany